Letts

Framework FOCUS

Maths

Dictionary

11-14

Gillian Rich

Published by Letts Educational
The Chiswick Centre
414 Chiswick High Road
London W4 5TF
tel: 020 89963333
fax: 020 87428390
email: mail@lettsed.co.uk
website: www.letts-education.com

Letts Educational Limited is a division of Granada Learning Limited, part of
Granada plc.

First published 2002
Reprinted 2002

ISBN 1 84085 697 1

British Library Cataloguing in Publication Data

A catalogue record for this book is available from the British Library.

Acknowledgements
Photography: Paul Mulcahy Photography
Cover photograph: Sheila Terry/Science Photo Library

Illustrations
Ken Vail Graphic Design, Cambridge

Commissioned by Helen Clark

Project management by Vicky Butt

Editing by Kim Richardson

Design by Ken Vail Graphic Design, Cambridge

Production by PDQ

Printed and bound in the UK by Ashford Colour Press

How to use this dictionary

The Letts Maths Dictionary is aimed at Key Stage 3 Maths students, although other students will also benefit from its clear explanations and interesting detail. Text and layout have been designed to make the dictionary easy to use, including many features to help you understand as much about the words as about the mathematical ideas. This will make you more confident when you meet the words in your reading, and when you use them in your writing. These features are described in the examples below.

Entry word, or headword
The main form of the word. Unusual forms of the word are given in brackets after the headword.

Definition
The meaning of the word. This is kept as clear and concise as possible.

Information
Background information on the word and its use.

penny (plural: pence) [p]
noun   1 2 3 4
Penny is a unit of money in the UK, as well as many other currencies.
🔢 100 pence = £1
ℹ️ Before decimal currency was introduced in the UK in 1971, 12 pence (12d) = 1 shilling (1s), and 20 shillings = £1. For some years after 1971, decimal pence were known as new pence.
➡ pound

Symbol
Any symbol or abbreviation of the headword is given in square brackets.

Example
An example of the headword being used in a sentence.

Related words
Other entries in the dictionary related to the headword.

square root [√]   1 2 3 4
noun
The square root of a certain number is the number that, when squared, gives that certain number.
🔢 The square root of 49 is 7.
➡ cube root, root

Topic area
The area in Maths in which the word is most commonly used (see full list on the next page).

Part of speech
This tells you the job that the word does in a sentence.

Other meanings
If a word has more than one meaning, each meaning is numbered and each is followed by its own example.

credit   1 2 3 4
1 noun
Credit is a loan given to purchase something.
🔢 How much interest needs to be paid on a credit of £500 at 5%?
2 verb
To credit a bank account is to pay money into that account.
🔢 They credited her account with £35.

Pronunciation
How to say the word.

Möbius strip (*mer*-*be*-*us*)
noun
A Möbius strip is a flat strip which is twisted halfway and the ends joined together. It has only one side and only one edge.
🔵

Illustration
Visual examples rather than verbal examples are given where appropriate.

🔴 The Möbius strip was invented in the 19th century by August Möbius, a German mathematician.

lozenge
➡ rhombus

Cross reference entry
Go to another entry instead.

Topic areas

Each entry is labelled with an icon which tells you its topic area:

1234   Numbers and the number system

+√x̄÷   Calculations

$x = a^2$ Algebra

▭   Shape, space and measures

⊞   Handling data

? ? ? ?   Applying maths and general vocabulary

✍   ICT

Aa

abacus (*ab*-a-kus)   **1 2 3 4**
noun

An abacus is a calculating frame with beads threaded on parallel wires or rods.

eg

ℹ️ In some countries abacuses are still used for addition, subtraction, multiplication and division. If you are skilful, you can calculate results more quickly with an abacus than with an electronic calculator.

absolute   **1 2 3 4**
adjective

The absolute value of a number is its numerical value, ignoring its sign.

eg $3 \times -4 \times 5 = -60$ but the absolute value equals 60.

accelerate (*ak-sel-er-ayt*)
verb

To accelerate means to increase speed.

eg When a moving object accelerates, its velocity increases relative to time.

accurate   **? ? ? ?**
adjective

To be accurate is to be exact or correct.

eg The test question asked for an accurate measurement.

➡️ **exact**

acre (*ay*-ker)
noun

An acre is a measurement of area in the imperial system: 1 acre = 4840 square yards.

eg The farm covered an area of 40 acres.

ℹ️ The use of acres as a measurement dates back to medieval times. An acre was the amount of land a yoke of oxen could plough in one day.

acute angle
noun

An acute angle lies between 0° and 90°.

eg

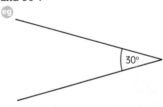

➡️ **angle, obtuse angle, reflex angle**

acute triangle
noun

An acute triangle is a triangle with 3 acute angles.

eg

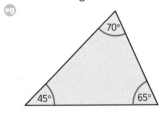

➡️ **obtuse triangle**

A
B
C
D
E
F
G
H
I
J
K
L
M
N
O
P
Q
R
S
T
U
V
W
X
Y
Z

add [+]
verb   +√x−÷

To add things, you join or connect them to each other.

eg Ramesh added 16 and 42.

addition [+] (ad-**i**-shun)
noun   +√x−÷

When you combine quantities together, you are doing addition.

eg Errol found the total by doing an addition sum.

adjacent (a-**jay**-sent)
preposition

Adjacent means near or next to something.

eg The side of the triangle adjacent to the right angle measured 15 cm.

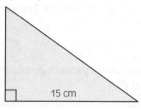

15 cm

➡ near

adjust
verb   1 2 3 4

To adjust something is to alter or adapt it.

eg Alisha decided to adjust her diagram by changing the scale.

algebra
noun $x = a^2$

Algebra is the part of mathematics which uses letters to stand for numbers.

eg In algebra, x and y often stand for different numbers.

ℹ Modern algebra was developed by a French mathematician, François Viète, in the 16th century. He was the first person to use letters for unknown quantities.

algebraic expression
noun $x = a^2$

An algebraic expression is a statement which uses letters as well as numbers.

eg Sam was asked to simplify the algebraic expression $2d + 7 - 3d + 4$.
➡ equation

alternate angle
noun

Alternate angles are formed when two or more lines are cut by a transversal. The angles are equal if the lines are parallel.

eg

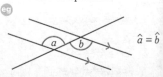

$\hat{a} = \hat{b}$

altitude
noun

The altitude of an object is its height, especially above sea level

eg The house stood at an altitude of 43 m.

a.m.
noun

The abbreviation for 'ante meridiem' is a.m., which means before midday or noon.

eg The train left the station at 11.45 a.m.
➡ p.m.

amount
+√x−÷

1 noun

The amount is the total quantity.

eg After everyone had paid for their school trip, the teacher banked the full amount.

2 verb

If something amounts to a final value it means it adds up to that value.

eg What will the final bill amount to after VAT is calculated?

analogue clock (***an**-a-log clok*)
noun
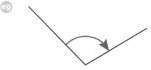

An analogue clock measures time using hands moving around a dial.

➡ **clock, digital clock**

analyse (***an**-a-lize*)
verb

To analyse something is to investigate or examine it by breaking it down into parts.

Claire needed to analyse the data before coming to any conclusions.

analysis (*an-**a**-li-sis*)
noun

An analysis is when something is broken down into separate parts for examination.

Sheena knew that the analysis of the data would help her investigation.

'and' rule
noun

The 'and' rule refers to the probability of a combination of independent events, and is expressed as
$P(A \text{ and } B) = P(A) \times P(B)$.

When 2 dice are rolled, use the 'and' rule to find the probability of getting a 3 and a 4.

➡ **independent, mutually exclusive, 'or' rule**

angle
noun

The amount of turning, measured in degrees, is an angle.

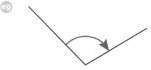

➡ **acute angle, obtuse angle, reflex angle**

angle of depression
noun

The angle of depression is the amount of turning from the horizontal down to a second line.

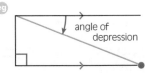

angle of depression

angle of elevation
noun

The angle of elevation is the amount of turning from the horizontal up to a second line.

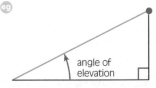

angle of elevation

angles at a point
noun

Angles meeting at the same place and adding up to 360° are called angles at a point.

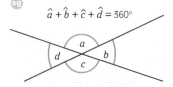

$$\hat{a} + \hat{b} + \hat{c} + \hat{d} = 360°$$

angles on a straight line – arithmetic

B
C
D
E
F
G
H
I
J
K
L
M
N
O
P
Q
R
S
T
U
V
W
X
Y
Z

angles on a straight line
noun

Two or more angles, at the same point on a straight line and adding up to 180°, are called angles on a straight line.

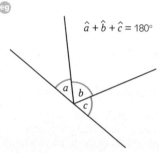

$$\hat{a} + \hat{b} + \hat{c} = 180°$$

➡ supplementary angle

annual
adjective

An annual event is one that occurs once a year.

The annual rainfall was illustrated using a bar chart.

answer
noun

? ? ? ?

An answer to a question is the solution.

Write the solution to the equation on the answer line.

anticlockwise
adjective, adverb

When something turns anticlockwise, it rotates in the opposite direction to the hands of a clock.

The triangle was rotated through 90° anticlockwise about one vertex.

➡ clockwise

apex
noun

An apex is the vertex at the highest point of a triangle, cone or pyramid.

apex

approximate
adjective

1 2 3 4

Approximate means nearly correct but not exact.

Give an approximate answer to the nearest whole number.

approximately equal to
[≈]

1 2 3 4

Approximately equal to means nearly, but not quite, exact.

2 kilometres is approximately equal to 1 mile.

arc
noun

An arc is a curved line, which can be part of a circumference.

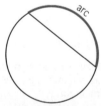
arc

area
noun

The surface of a shape or object is called its area.

The area of the room measured 4.5 square metres.

➡ square metre, surface area

arithmetic
noun

1 2 3 4

Arithmetic is the part of mathematics that deals with calculating numbers or counting.

Brian used an abacus to do his arithmetic.

arithmetic mean
➡ mean

arithmetic sequence
noun $x = a^2$

An arithmetic sequence is one where terms are found by adding or subtracting a constant amount.

eg Continue the arithmetic sequence 1, 3, 5, 7…

arrange
1234
verb

To arrange something is to plan, organise or manage it.

eg William arranged the blocks to form a large cuboid.

arrangement
1234
noun

Something which is arranged is called an arrangement.

eg The arrangement of the numbers formed a magic square.

array
1234
noun

A regular arrangement, in the form of rows and columns, is called an array.

eg

1	2	3
4	5	6
7	8	9

arrow
noun

A sign in the shape of an arrow is often used as a pointer in mathematics.

eg Mark the position of the house on the map using an arrow.

arrowhead
noun

An arrowhead is a quadrilateral with 2 pairs of equal adjacent sides and 1 interior corner angle greater than 180°. It is sometimes called a dart.

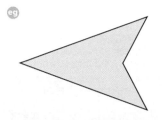

ascending
1234
adjective

Ascending means going up in order from smallest to largest.

eg Write 0.53, 0.526, 0.555, 0.54 in ascending order.

associative (*a-so-si-a-tiv*)
adjective $+\sqrt{x^-\div}$

An operation is associative if it is independent of the grouping of the numbers.

eg Addition and multiplication are associative, subtraction and division are not: $11 \times (2 \times 5) = (11 \times 2) \times 5 = 110$.
➡ commutative

at [@]
preposition $+\sqrt{x^-\div}$

At, in the form of the symbol @, is used to refer to the rate or cost of something.

eg What is the cost of 10 pens at 30p each?

average
1 adjective

An average value is the normal or standard amount or value.

eg What is the average income of a teacher in this country?

2 noun

Average is also a general term to refer to the mean, median or mode.

eg The mean is used as the average when all data is used and an exact calculation is needed.
➡ mean, median, mode

average speed
noun

The average speed is found by dividing the total distance travelled by the total journey time.

eg What is the average speed of a car travelling a distance of 150 km in 2 hrs?

axis (plural: axes)
$$x = a^2$$
noun

An axis is the horizontal or vertical line on a graph from which coordinates are measured.

eg

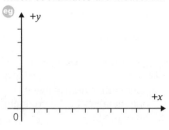

axis of rotation symmetry
noun

An axis of rotation symmetry is a line about which a shape can be rotated and remain the same.

eg

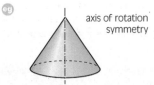

axis of rotation symmetry

➡ rotation symmetry

axis of symmetry
noun

An axis of symmetry is a line about which a shape can be rotated or reflected and remain the same.

eg

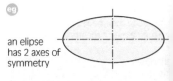

an elipse has 2 axes of symmetry

➡ symmetry

Bb

Babylonian number system
1 2 3 4

noun

The Babylonian number system was a number system based on 60.

eg Our measurement of time and angles is based on 60 because of the Babylonian number system.

balance
1 noun

Balance is another name for a weighing machine or scales.

eg

2 verb $x = a^2$

To balance is keep two items matched.

eg Simon subtracted 7 from both sides of the equation to keep it balanced.

bar
noun

A bar is a horizontal or vertical column used in a bar chart.

eg Plot the bar at the correct height to represent the frequency.

bar chart
noun

A bar chart is a chart that uses bars of equal width to represent statistics.

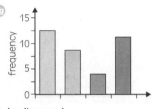

➡ **bar-line graph**

bargain
1 2 3 4

noun

A bargain is a good deal.

eg Did you get a bargain when you bought 3 books for the price of 2?

bar-line graph
noun

A bar-line graph is a bar chart where the bars appear as lines.

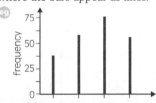

➡ **bar chart**

base
1 noun

The base of a figure is the bottom line or surface.

eg

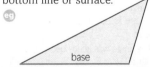

base

2 noun **1 2 3 4**

A number base is the number on which a counting system depends.

eg The base of the decimal system is the number 10.

base angle
noun

The base angles of a figure are the angles at the ends of the base.

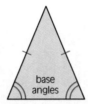

base angles

basic
adjective

Basic means elementary or fundamental.

eg The survey began with some basic questions.

bearing
noun

A bearing, measured in degrees, gives the direction of a travelling object. The bearing is measured clockwise from the north line and is given as a 3-figure angle.

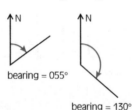

bearing = 055°

bearing = 130°

because
conjunction

Because is used to introduce a reason.

eg The angle was acute because it was less than 90°.

belong
verb

Something that belongs is part of or connected to something else.

eg The rhombus belongs to the set of quadrilaterals.

beside
preposition

To be beside something is to be close to it.

eg Write the label beside the figure.
➡ adjacent

best estimate
noun

$+\sqrt{x}^-\div$

The best estimate is the estimate giving the best approximation.

eg The best estimate of the value of $701 \div 19$ is 35.

best fit
➡ line of best fit

between
preposition

If something is between two other things, it is situated in the space that separates them.

eg Pick a number between 1 and 10.

beyond
preposition

To be beyond something is to be further than it or past it.

eg The batsman hit the ball beyond the boundary.

biased
adjective

Biased means having a tendency towards some value or away from the normal.

eg The dice was obviously biased, as it kept showing 6.
➡ unbiased

BIDMAS
noun

$x = a^2$

BIDMAS is an acronym which helps you remember the order of operation.

eg BIDMAS stands for **B**rackets (**I**ndices & roots) (**D**ivision & **M**ultiplication) (**A**ddition & **S**ubtraction).

🔍 BIDMAS sometimes occurs as BODMAS where 'O' is 'orders or powers'.

bill
1 2 3 4
noun

A bill is an account of what is owed.

eg The plumber presented the bill after he had fitted the washing machine.

billion
1 2 3 4
noun

A billion is a thousand million.

eg 1 billion = 1 000 000 000 or 10^9

i In the UK, a billion was originally a million million (1 000 000 000 000).

birthday
noun

A person's birthday is the anniversary of their birth.

eg On your birthday you become 1 year older.

bisect
verb

To bisect is to divide a line, angle or area exactly in half.

eg Bisect the 8 cm line into 2 equal parts of 4 cm.

bisector
noun

A bisector is a line which divides a line, angle or area exactly in half.

eg

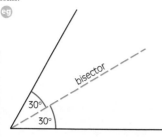

➡ **perpendicular bisector**

block
noun

A block is an object which is usually the shape of a brick or cuboid.

block graph
noun

A block graph is a chart using a block to represent one observation.

eg

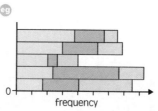

board
noun

A board is the square base used for a game such as chess or draughts.

eg The chess board was 8 squares by 8 squares, making 64 squares.

BODMAS
➡ **BIDMAS**

borrow
1 2 3 4
verb

To borrow is to take a loan of something from somebody else.

eg When Holly borrowed money from the building society, she had to pay interest of 7.5%.

➡ **lend, loan**

bound
noun

A bound is a limit.

> The graph was drawn within the bounds of $-3 < x < +3$.

bracket
noun

$x = a^2$

Brackets are symbols used to show items that are treated together.

> In arithmetic and algebra, operations within brackets are given priority: $4 \times (5 + 3) = 32$, whereas $4 \times 5 + 3 = 23$.

branch
noun

A branch is an arm or subsection of a probability tree.

>

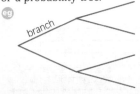

branch

breadth
noun

Breadth is another name for width. It is the distance across, from side to side.

> What is the breadth of a rectangle which has area 64 cm² and length 16 cm?

➡ length, width

broad
adjective

Broad means wide or wide ranging.

> The data collected need to be over a broad range, covering all values.

➡ narrow, wide

budget
noun

1234

Your budget is the estimate of how much money you will spend.

> The budget for the school outing was approximately £350.

building society
noun

1234

A building society is an organisation in which people can invest money, and which lends people money for large purchases.

> When they wanted to buy a house, Dennis and Joan applied to the building society for a mortgage.

Building societies were originally called friendly societies. Many banks now act as building societies. In the USA similar societies are called savings and loan associations.

➡ loan, mortgage

business
noun

1234

A business is a trade or occupation.

> The business needed to make a profit to make sure it could keep trading.

buy
verb

1234

To buy is to purchase – to pay money for an item or service.

> How many oranges, at 6p each, can you buy for £1?

byte (*bite*)
noun

A byte is an amount of computer memory needed to store a unit of information.

> Computer memory size is measured in multiples of bytes.

Like all computer vocabulary, byte is a relatively new word. It was created in the 20th century as a combination of 'bit' and 'bite'.

➡ kilobyte, megabyte

Cc

C 1 2 3 4

C is the symbol which stands for 100 in the Roman number system.

eg C + VI + IV = CX

➡ **Roman numerals**

calculate +√x−÷
verb

To calculate is to use numbers to work out values.

eg Calculate the third angle of an isosceles triangle having base angles of 40°.

ℹ Calculate comes from the Latin word *calculus*, meaning a pebble used as a counter.

calculator +√x−÷
noun

A calculator is a machine used for calculating.

eg

calendar
noun

A calendar is a method of ordering the year.

eg Tom used the calendar to add up the days he attended school.

➡ **year**

calendar year
➡ **year**

cancel 1 2 3 4
verb

To cancel is to simplify a fraction by dividing numerator and denominator by a common factor.

eg Simplify these fractions by cancelling: $\frac{26}{52}, \frac{22}{28}, \frac{45}{60}$.

➡ **simplify**

capacity (*ka-pa-si-tee*)
noun

Capacity is the amount of space in a container or the amount of liquid it will hold.

eg In the metric system capacity is measured in litres; in the imperial system it is measured in pints and gallons.

➡ **volume**

cardinal number 1 2 3 4
noun

A cardinal number is a number that shows quantity but not order.

eg The cardinal numbers are 1, 2, 3, 4...

➡ **ordinal number**

cash 1 2 3 4
noun

Cash is money in the form of coins or notes.

eg The cash was put in the safe overnight.

category
noun

A category is a class or group.

eg Sort the data into the different categories.

cause
noun

A cause is a reason for something happening.

eg The cause of the low average temperature was a period of cold weather.

CD-ROM
noun

CD-ROM (Compact Disk Read-Only Memory) is a method of storing information on a disk that can be used on a computer.

eg The encyclopedia was accessed by using the CD-ROM.

cease
verb

To cease is to stop.

eg After some time the ball ceased bouncing.

Celsius (*sell-see-us*)
adjective

Celsius refers to the Celsius scale, used for measuring temperature.

eg On the Celsius scale, the freezing point of water is 0°C and the boiling point of water is 100°C.

i The Celsius scale was named after a Swedish astronomer, Anders Celsius. His first scale, developed in 1742, was the reverse of the modern scale: he used 100° for freezing point and 0° for boiling point.

➡ **centigrade, Fahrenheit**

census
noun

A census is a count of a population.

eg A census is carried out in the UK every 10 years.

cent
noun

1 2 3 4

A cent is a unit of money used in some countries. Its value is $\frac{1}{100}$ of the main currency unit.

eg In the USA 100 cents equal 1 dollar.

centi-
prefix

Centi- is a prefix meaning $\frac{1}{100}$.

centigrade
adjective

Centigrade refers to the centigrade scale, used for measuring temperature. Centigrade is now usually replaced by Celsius.

eg On the centigrade scale, the freezing point of water is 0°C and the boiling point of water is 100°C.

➡ **Celsius, Fahrenheit**

centigram [cgm]
noun

A centigram is a measurement of weight equal to $\frac{1}{100}$ gram.

➡ **gram**

centilitre [cl]
noun

A centilitre is a measurement of capacity equal to $\frac{1}{100}$ litre.

➡ **litre**

centimetre [cm]
noun

A centimetre is a measurement of length equal to $\frac{1}{100}$ metre.

➡ **metre**

centre
1 noun

The centre is a fixed point at the middle of a circle or a sphere.

eg

centre

2 noun

In general, the centre is the middle point of something.

eg Stand in the centre of the room.

centre of enlargement
noun

A centre of enlargement is the point from which the enlargement happens.

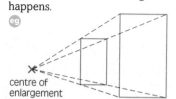

centre of enlargement

➡ **enlargement**

centre of rotation
noun

A centre of rotation is the point around which a shape can rotate.

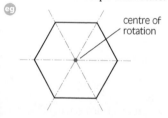

centre of rotation

century

1 noun

A period of 100 years is called a century.

eg Did the 20th century end on 31 December 2000 or on 31 December 1999?

2 noun

A batsman who scores 100 runs in a cricket match has made a century.

eg Hussain scored two centuries in three matches.

certain
adjective

To be certain about something is to be absolutely sure that it will happen.

eg If an event is certain, its probability equals 1.

➡ **uncertain**

chain

1 noun

A chain is a measurement of length in the imperial system.

eg 1 chain = 22 yards or 66 feet (20.13 metres)

2 noun

A length of joined metal links or rings is called a chain.

chance
noun

Chance means probability.

eg What is the chance that it may rain today?

(1) If an event is impossible, there is **no chance** and probability = 0.

(2) If an event is unlikely to happen, there is a **poor chance** and $0 < $ probability $ < 0.5$.

(3) If an event has an **even chance**, probability = 0.5.

(4) If an event is likely to happen, there is a **good chance** and $0.5 < $ probability $ < 1$.

➡ **fifty-fifty, probability**

change

1 noun

A change is an alteration.

eg Make a change to the figure by adding a side.

2 verb

To change is to alter.

eg Change the parallelogram by enlarging it by a scale factor of 3.

charge   **1234**

1 noun

A charge is the price of an item or service.

eg The entry charge is £3.50 per person.

2 verb

To charge for an item or service means to ask someone to pay for it.

eg How much will a group of 12 people be charged?

➡ cost price

chart

noun

A chart is an illustration of statistical data.

eg

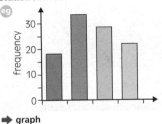

➡ graph

cheap

1234

adjective

If an item is cheap, it costs a small amount.

eg Everything in the sale is very cheap.

➡ dear

check

? ? ? ?

verb

To check something means to inspect it to see if it is correct.

eg When you have finished, check the results.

cheque (check)

1234

noun

A cheque is a written order to a bank to pay money.

eg The customer wrote a cheque to pay for the goods.

choice

noun

A choice is an option or selection.

eg It is your choice whether you use pen or pencil.

choose

verb

To choose is to make a selection.

eg Choose the prime numbers from the following list: 2, 4, 9, 13, 21 and 23.

chord

noun

A straight line joining the ends of an arc of a circle is called a chord.

eg

➡ diameter, segment

chronological

adjective

When items are arranged in order of time, they are in chronological order.

eg List the major events of your life in chronological order.

circle

noun

A circle is a shape with every point on its edge at a fixed distance from the centre.

eg Calculate the circumference of the circle.

➡ centre, chord, circumference, pi, radius, sphere

circumcentre

noun

A circumcentre is the centre of a circumcircle.

eg The centre of the circle passing through every vertex of a regular pentagon is the circumcentre.

circumcircle
noun

A circumcircle is the circle drawn around the outside of any regular polygon so that it touches every vertex.

eg

circumference
noun

The circumference is the edge or perimeter of a curved shape such as a circle or ellipse; it is also the length of that edge.

eg The circumference of a circle of radius r and diameter d can be found using the formula $C = 2\pi r$ or πd.

➡ **circle, perimeter, pi**

circumscribe
verb

If a figure can have another figure drawn around the outside so that it touches every vertex, the second figure circumscribes the first figure.

eg

class
noun

A class of data is a group of that data.

eg Vicky separated her data into different classes.

➡ **grouped data**

classify
verb

To classify data is to arrange it into groups or classes.

eg When data collection is completed, you must classify it.

class interval
noun

When collecting data, each class or group is bound by the limits of the class interval.

eg The data about age was sorted into class intervals of 5 years.

clear
verb

$+\sqrt{x}-\div$

To clear a calculator memory means to empty it.

eg When you have finished the calculation, clear the memory and the display.

climate
noun

The climate of a place is the average weather conditions of that place.

eg Referring to the bar chart on weather in the local guidebook, Jared wrote about the climate of his home town.

clinometer (*klin-**om**-it-er*)
noun

A clinometer is a hand-held instrument for measuring the angle of elevation or depression.

eg

clock
noun

A clock is an instrument for measuring time.

eg A clock cannot be worn, unlike a watch.

i A 12-hour clock face is divided into 12 sections, each of 5 minutes. It has 2 hands, fixed at the centre. One hand is shorter than the other: it measures the hours, completing a revolution in 12 hours. The longer hand measures minutes, completing a revolution in 60 minutes or 1 hour.

➡ **analogue clock, digital clock, time**

clockwise
adjective, adverb

The hands moving round a clock face move in a clockwise direction.

eg

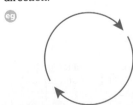

➡ **anticlockwise**

coaxial
adjective

Two objects or shapes with the same axis are coaxial.

eg

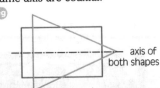

axis of both shapes

code
$x = a^2$

noun

A code is information translated into words, letters or numbers.

eg A B C D E F G …
      1 2 3 4 5 6 7 …

This code translates BED into 254.

coefficient
$x = a^2$

noun

A coefficient is a number or letter multiplying an algebraic term.

eg In $ax^2 + bx + c$ the coefficient of x^2 is a, and the coefficient of x is b.

collect
? ? ? ?

verb

To collect is to gather together.

eg Henry had to collect all the data for his investigation.

collect like terms
➡ **like terms**

column
1 2 3 4

noun

A column is a vertical arrangement.

eg Anya added up all the numbers in the column.

➡ **row**

column graph
noun

A bar graph in which the bars are presented vertically is also called a column graph.

eg

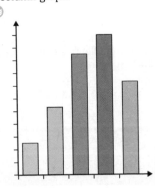

combination
noun

A combination of items is a mixture or compound of those items.

eg Find the probability of a combination of the 2 events.

command
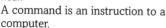
noun

A command is an instruction to a computer.

eg A command consists of a line of code telling the computer to carry out a task.

common
? ? ? ?
adjective

If something is common it is shared by 2 or more things.

eg When a diagonal is drawn in a rectangle, dividing it into 2 triangles, the diagonal forms a side common to both triangles.

➡ **common factor**

common factor
1 2 3 4
noun

A common factor is one which is the same for two or more numbers.

eg Which are the common factors of 15, 39 and 45?

➡ **cancel, highest common factor, take out common factors**

common fraction
1 2 3 4
noun

A common fraction (also known as a simple fraction or a vulgar fraction) has a numerator and a denominator that are both integers.

eg $\frac{2}{3}$, $\frac{15}{16}$ and $\frac{23}{33}$ are all common fractions.

common multiple
1 2 3 4
noun

If 2 or more numbers have some of the same multiples, they are known as common multiples.

eg Use multiplication tables to find some common multiples of 2, 4 and 6.

➡ **common, lowest common multiple, multiple**

commutative
(*kom-**mew**-ta-tiv*)
adjective
$+\sqrt{x}^{-}\div$

An operation is commutative if it is independent of the order of the numbers.

eg Addition and multiplication are commutative, subtraction and division are not: $11 + 2 + 5 = 5 + 2 + 11 = 18$.

➡ **associative**

compare
? ? ? ?
verb

To compare is to see how similar things are.

eg Look at the triangles and compare the angles and corresponding sides.

compass
noun

A compass is an instrument used to find the direction of north.

➡ **bearing**

compasses
noun

Compasses are an instrument used for drawing circles.

compensation
1 2 3 4
noun

Compensation is used when rounding or adjusting answers.

A B **C** D E F G H I J K L M N O P Q R S T U V W X Y Z

eg 23 + 19 is rounded to 23 + 20 to make it easier to calculate. The result of 43 needs a compensation of −1 to give the correct answer of 42.

complement
noun $+\sqrt{x^-\div}$

A number and its complement have a given total.

eg When considering complements of 100, 57 has a complement of 43.

complementary angle
noun

Angles that add up to 90° are called complementary angles.

eg Which angle is complementary to 63°? The complementary angle = 90° − 63° = 27°.

➡ **supplementary angle**

complete
? ? ? ?

1 adjective

A complete item has all its parts.

eg The complete pie chart had labels for each sector.

2 verb

To complete is to finish.

eg Ben made sure he completed the scatter graph before he drew the line of best fit.

complex
? ? ? ?

adjective

If a task is complex it is not simple but complicated.

eg The task had to be broken down into smaller tasks, as it was so complex.

compound
adjective

A compound figure is a mixture of more than one shape.

eg

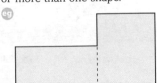

compound interest
noun **1 2 3 4**

Compound interest is interest calculated by adding the interest each year before calculating the new amount.

eg Find the compound interest earned after 3 years on an investment of £500 at 6% per annum.

➡ **interest, simple interest**

computer
noun

A computer is a machine for processing data.

eg The monitor and printer are all examples of computer hardware; the program code, which enables the computer to perform set tasks, is an example of software.

The first mechanical computer was designed by Charles Babbage in 1835. The first electronic computer was built by Thomas Flowers and Alan Turing in 1943.

concave
adjective

A surface curving inwards is concave.

eg

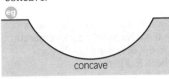

concave

➡ **convex**

concentric
adjective

One or more curved shapes that have the same centre are concentric.

eg

conclude
? ? ? ?

verb

To conclude is to finish.

(eg) Michael concluded the calculation by underlining the answer.

conclusion
? ? ? ?

noun

A conclusion sums up the results of an investigation.

(eg) Make the final paragraph of your coursework the conclusion.

congruent

adjective

Congruent shapes have equal angles and sides.

(eg) David compared the sides and angles of the triangles and found that they were congruent.

connect

verb

To connect is to join.

(eg) Connect a rectangle to a semicircle to make a compound shape.

consecutive
1 2 3 4

adjective

Consecutive means continuous or following in order.

(eg) The numbers 1, 2, 3… are consecutive.

consist
? ? ? ?

verb

To consist of different parts means to be made up of those parts.

(eg) Ann built a model that consisted of several cuboids.

constant

noun

An unvarying number or quantity is a constant.

(eg) $\pi = 3.142$ is a constant that is used in circle formulae.

construct

verb

To construct a figure is to draw it accurately.

(eg) Construct a triangle with one side of 6 cm, another side of 7.5 cm and an angle of 60° between them.

construction line

noun

A construction line is a line used when constructing accurate diagrams.

(eg) When you draw the diagram, remember to leave in your construction lines and arcs.

consumer

noun

A consumer is a user of products.

(eg) The supermarket conducted a survey to find out how consumers wanted to use its store.

contain

verb

To contain is to hold or include.

(eg) The tank contained 500 litres of water.

contents

noun

The contents are everything that is inside a container.

(eg) The contents of the can weighed 450 grams.

continue
1 2 3 4

verb

To continue is to carry on.

(eg) A recurring decimal has digits that continue without limit, like 0.3333333…

continuous data

noun

Continuous data are data arranged in groups with no gaps.

(eg) Measuring the height of plants will give continuous data over a range of values.

➡ **discrete data**

convention ? ? ? ?
noun

A convention is a usual or common rule for writing down mathematics.

eg It is a convention to use capital letters to label geometrical shapes.

convert 1 2 3 4
verb

To convert something is to change it from one form to another.

eg How many dollars will you receive when you convert £10 into dollars at a rate of $1.40 for £1?

convex
adjective

A surface curving outwards is convex.

eg

convex

➡ concave

coordinate $x = a^2$
noun

Coordinates are numbers giving the position of a point on a graph or grid. Also called coordinate pairs (or coordinate points), they are given in the form (x, y).

eg

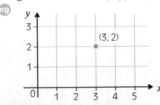
(3, 2)

➡ plot, x-coordinate, y-coordinate, z-coordinate

corner
noun

A corner is a point where 2 or more lines meet.

eg A square has 4 corners.

➡ vertex

correct
1 adjective ? ? ? ?

Correct is the same as right.

eg Her homework was all correct.

2 verb 1 2 3 4

To correct is to improve or adjust.

eg 36.25 corrected to 3 significant figures is 36.3.

correlation
noun

Correlation is the connection between 2 variables, which can be found by drawing a scatter graph.

eg

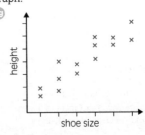
height
shoe size

➡ line of best fit, scatter graph

correspond
verb

To correspond means to match.

eg To find out if triangles are similar, you need to compare the ratio of corresponding sides, i.e. sides opposite the same angle.

corresponding angle
noun

Corresponding angles are formed when a transversal cuts across 2 or more lines. They are equal when the lines are parallel.

eg

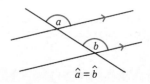
a
b
$\hat{a} = \hat{b}$

cosine [cos]
noun

In a right-angled triangle, the cosine is the ratio of the length of the side adjacent to the given angle to the hypotenuse.

eg

$$\cos X = \frac{\text{adjacent}}{\text{hypotenuse}}$$

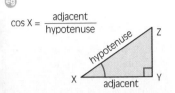

➡ **sine, tangent, trigonometry**

cost price
noun

The price of an item when it is manufactured is the cost price, which is sometimes known as the wholesale price.

eg The cost price of the trousers was half the sale price.

➡ **sale price**

count
verb

To count is to number.

eg Count the steps you need to walk round the playground.

counter
noun

A counter is a small round piece of plastic or metal used to play a game.

eg How many counters are left?

counter-example
noun

A counter-example is an example which shows the opposite of the original statement.

eg All prime numbers are odd, but the counter-example is 2, which is a prime number and even.

couple
noun

A couple is two of a kind.

eg An isosceles triangle has a couple of equal sides and a couple of equal angles.
➡ **pair**

create
verb

$x = a^2$

To create is to make or form.

eg Create an equation by thinking of a number, multiplying by 4 and then subtracting 5, giving the result 23.

credit
1 noun

Credit is a loan given to purchase something.

eg How much interest needs to be paid on a credit of £500 at 5%?
2 verb

To credit a bank account is to pay money into that account.

eg They credited her account with £35.

cross
noun

A cross is a shape formed by two lines, often at right angles.
eg Draw 2 different crosses.

cross-section
noun

A cross-section of a solid figure is a slice cut through at an angle of 90° to its axis.

eg The cross-section of a cylinder is a circle.
➡ **section**

cube
1 noun

A cube is a solid figure with 6 square faces.

➡ **cuboid, square**

2 noun **1 2 3 4**

The product of 3 equal numbers is the cube of that number, or that number cubed.

eg 4 cubed (4^3) is $4 \times 4 \times 4$.

cube root [$\sqrt[3]{}$] **1 2 3 4**
noun

The cube root of a certain number is the number which, when cubed, will give that certain number.

eg $\sqrt[3]{27} = 3$ because $3 \times 3 \times 3 = 27$.

➡ **root, square root**

cubic centimetre [**cc**]
noun

A cubic centimetre is a measurement of volume equal to 1 centimetre cubed.

eg

cubic curve $x = a^2$
noun

A cubic curve is a graph drawn from a cubic equation or function.

eg

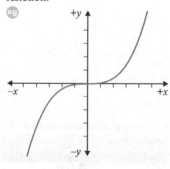

cubic equation $x = a^2$
noun

A cubic equation is an equation containing unknowns to a maximum power of 3.

eg $y = x^3 + 2x^2 - 5x$ is a cubic equation.

cubic expression $x = a^2$
noun

A cubic expression is an expression containing unknowns to a maximum power of 3.

eg $x^3 - 4x^2 + 2x + 3$ is a cubic expression.

cubic function $x = a^2$
noun

A cubic function is a function containing unknowns to a maximum power of 3.

eg $f(x) \to x^3 - 4x^2 + 2x + 3$ is a cubic function.

cubic metre [m^3]
noun

A cubic metre is a measurement of volume equal to 1 metre cubed.

eg

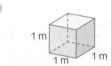

cubic millimetre [mm^3]
noun

A cubic millimetre is a measurement of volume equal to 1 millimetre cubed.

eg

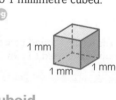

cuboid
noun

A cuboid is a solid shape that has 6 rectangular faces.

eg

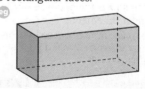

➡ **cube**

cumulative frequency
noun

Cumulative frequency is found by adding previous amounts to each group's frequency until the total frequency is found.

eg cumulative frequency table

mark	freq(f)	mark	cum freq
1–10	2	≤10	2
11–20	10	≤20	12
21–30	15	≤30	27
31–40	20	≤40	47
41–50	16	≤50	63
51–60	12	≤60	75

cumulative frequency graph
noun

A cumulative frequency graph is a graph drawn from a cumulative frequency table. Points are plotted at the upper limit of each class or group, then connected by a smooth curve.

eg

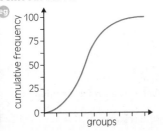

currency
noun

Currency is the money of a country.

eg From 2002 the currency of many European countries became the euro.

➡ **exchange rate**

cursor
noun

A cursor is a movable arrow or pointer on a computer monitor screen.

eg You need to type the word at the flashing cursor.

curve
noun

A curve is a line that is not straight.

eg A circle is a continuous curve, which is always the same distance from a given point.

cylinder
noun

A cylinder is a regular solid that has a cross-section of a circle.

eg

D d

D
1 2 3 4

D is the symbol which stands for 500 in the Roman number system.

eg M + D − CLX = MCCCXL

➡ **Roman numerals**

daily
adjective

Something done every day is done daily.

eg The newspaper is delivered daily except for Sunday.

➡ **day**

dart
➡ **arrowhead**

data
noun

Data are a collection of numbers or information.

eg Unprocessed information is called raw data.

database
noun

A database is a large amount of information stored in an organised way, often on a computer.

eg He set up a database of all the names and addresses.

data collection sheet
noun

A data collection sheet, or data log, is used to record observed data.

eg The data collection sheet showed the observations, tally and frequency.

data log
➡ **data collection sheet**

day
noun

A day is a 24-hour period that begins with midnight.

eg A day is the time it takes for the Earth to turn once on its axis.

➡ **daily**

dear
1 2 3 4
adjective

Something which is expensive is said to be dear.

eg She did not buy the trainers as they were too dear.

➡ **cheap, expensive**

debt (det)
1 2 3 4
noun

A debt is money owed to someone.

eg How much interest is due on the outstanding debt?

➡ **loan**

deca-
1 2 3 4
prefix

Deca- is a prefix meaning 10.

decade
noun

A decade is a time period of 10 years.

eg There are 10 decades in a century.

decagon
noun

A decagon is a 10-sided polygon.

eg

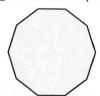

decahedron
noun

A decahedron is a 10-faced solid figure.

eg There is no regular decahedron.

deci-
1234
prefix

Deci- is a prefix meaning $\frac{1}{10}$.

decimal
1234
adjective

Decimal means relating to or using the base 10 or powers of 10.

eg The metric system is a decimal system.

decimal fraction
1234
noun

The part of a decimal number to the right of the decimal point is called the decimal fraction.

eg The decimal fraction 0.4321 is $\frac{4}{10} + \frac{3}{100} + \frac{2}{1000} + \frac{1}{10000} = \frac{4321}{10000}$.

decimal number
1234
noun

A decimal number is a number that only has digits between 0 and 9.

eg A decimal number is often just called a decimal.

decimal place
1234
noun

The position of a digit after the decimal point is known as its decimal place.

eg 3.456 has 3 decimal places.

decimal point
1234
noun

The decimal point is the dot used to divide the whole numbers and the decimal fraction.

eg Make sure to put the decimal point in the correct place.

i The use of decimal points varies around the world. In France, for example, a comma is used instead of a point.

deck
➡ pack

decomposition
$+\sqrt{x}-\div$
noun

Decomposition is a vertical method of subtraction.

eg Decomposition breaks down numbers in the top line, borrowing 10 from the next column.
➡ subtraction

decrease
$+\sqrt{x}-\div$

1 noun

A decrease is a reduction.

eg The Bank of England announced a decrease in interest rates.

2 verb

To decrease is to lessen or reduce.

eg Thomas decreased the size of the pentagon by a scale factor of $\frac{1}{3}$.
➡ increase

deduce
? ? ? ?
verb

To deduce means to draw a conclusion or make a deduction.

eg As the quadrilateral had 4 angles of 90° and 2 pairs of parallel equal sides, Misha deduced that it was a rectangle.

deduct
verb

To deduct is to subtract.

eg I need to deduct 20 from the total.

define
verb

When you state the meaning of something precisely, you define it.

eg Define the quadrilateral by determining its properties.

degree [°]
noun

1 A degree is a unit of measurement of temperature.

eg Water boils at 100 degrees Celsius and 212 degrees Fahrenheit.

2 A degree is also a unit of measurement of angles.

eg A complete revolution is 360 degrees.

degree of accuracy
noun **1 2 3 4**

A degree of accuracy is the level of approximation that you make, for example to the nearest £, cm, thousand.

eg The figure 68 700 is given to a degree of accuracy of the nearest hundred.

delta
noun

A delta shape is a quadrilateral which is almost a triangle. It has 1 pair of equal sides with an included obtuse angle. Its other pair of equal sides almost form a straight line.

eg

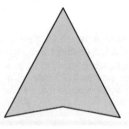

The shape is named after the Greek letter Δ (D), called delta.

denominator
noun **1 2 3 4**

The bottom of a fraction, below the line, is called the denominator.

eg In the fraction $\frac{2}{3}$, the denominator is 3.

➡ fraction, numerator

density
noun

The density of an object is its mass per unit volume.

eg Calculate the density of concrete in a block (10 × 15 × 20 cm) with mass 10 kg.

depend
verb

To depend on someone or something is to rely on them.

eg When drawing a graph, one variable y changes with the other variable x. Therefore y depends on x.

deposit
1 2 3 4

1 noun

Savings paid into a bank or building society is a deposit, on which interest is earned.

eg Fran made a deposit of £500.

2 noun

A deposit is also an amount of money paid as a first instalment against the full price of the item.

eg The freezer cost £275. Greg put down a deposit of £50 and paid the balance over 6 months.

3 verb

When money is put into a bank account, it is deposited.

eg Mai deposited her wages into her bank account.

➡ bank, interest

depression
noun

A depression is a lower area of ground.

eg The monument was in a depression at the bottom of the hill.

➡ angle of depression

depth
noun

When something is at a certain depth, it is that distance under the ground or the sea.

eg The shipwreck was found at a depth of 15 fathoms.

derived property

noun

A derived property is a feature that is not essential to a definition, but a consequence of it.

🔹 The fact that opposite sides of a parallelogram are equal in length is a derived property, not a definition.

descending

1 2 3 4

adjective

Descending means going down in order from largest to smallest.

🔹 Write 0.53, 0.526, 0.555, 0.54 in descending order.

➡ ascending

describe

? ? ? ?

verb

To describe something is to explain what it is by giving details.

🔹 A pentagon can be described as a 5-sided polygon.

design

1 noun

A design is a plan or model.

🔹 The model demonstrated the design of the new playground.

2 verb

To design something is to plan it.

🔹 Gillian designed the mosaic before she started collecting the pieces.

determine

verb

To determine something is to decide what is needed.

🔹 First determine which maths apparatus and data you need.

diagonal

noun

A line joining 2 non-adjacent corners or vertices of a shape or object is called a diagonal.

🔹

diagonal

diagram

noun

A diagram is a line drawing or illustration.

🔹 Remember to draw a rough diagram first.

diameter

noun

The chord that passes through the centre of a circle or sphere is called the diameter.

🔹

diameter

centre

➡ circle, semicircle

diamond

➡ rhombus

dice (singular: die)

noun

Dice are 6-sided cubes with a number from 1 to 6 marked on each face. The sum of the opposite sides is 7.

🔹

difference

+√x⁻÷

noun

The amount found between numbers when subtracting is the difference.

🔹 A pair of shoes costs £45. Find the difference between a discount of £10 in one shop and a 20% discount in another.

difference of 2 squares – discrete data

difference of 2 squares
noun $x = a^2$

The difference of 2 squares is the factorisation of $x^2 - y^2$ which equals $(x + y)(x - y)$.

eg An example of difference of 2 squares is $4a^2 - 25 = 4a^2 - 5^2 = (2a + 5)(2a - 5)$.

difference pattern
noun $x = a^2$

A difference pattern generates a sequence by using differences between consecutive terms.

eg There is a regular difference pattern in the sequence 1, 3, 5, 7…, so the next term is 9.

➡ **first/second difference**

digit
noun 1234

A digit is any of the 10 numerals from 0 to 9.

eg The value of the digit 7 in the number 1,275 is 70.

digital clock
noun

A digital clock is a clock which indicates the hours and minutes by digits rather than by hands on a dial.

eg

➡ **analogue clock**

dimension
noun

The number of coordinates needed to define the position or size of an object is its dimension.

eg A line has 1 dimension (length), a plane shape has 2 dimensions (length and breadth) and a solid has 3 dimensions (length, breadth and height).

➡ **three-dimensional, two-dimensional**

direct
verb

To direct something is to move it according to a given order.

eg In the treasure hunt, Dan followed the instructions which directed him to each clue.

directed number
noun 1234

A number is directed when it is given a sign + or −.

eg When using directed numbers, sign rules must be obeyed.

➡ **sign rule**

direct proportion
noun 1234

When numbers change in direct proportion, the ratio of one to the other remains the same.

eg If Georgia is paid £27.50 for 5 hours' work, her earnings for 8 hours' work will be in direct proportion of $\frac{8}{5}$.

➡ **inverse proportion**

disc
noun

A disc is a 2-dimensional circular shape.

eg A disc of radius 5 cm was cut out of a square piece of plastic of side 15 cm.

discount
noun $+\sqrt{x}-\div$

A discount is an amount subtracted from the original price of an item.

eg If the coat cost £50 after a discount of 25%, what was its original price?

discrete data
noun

Separate or distinct items or groups of data are known as discrete data.

eg Shoe sizes are discrete data.

➡ **continuous data**

disk
noun

A disk, or diskette, is a piece of computer equipment that carries data. It is fitted into a disk drive in the computer.

diskette
➡ disk

displacement
noun

A change of position is a displacement.

eg The displacement of an object is the distance moved in a direction described by a vector.

➡ translation, vector

display
verb +√x−÷

To display is to represent information visually.

eg The digital clock displayed a time of 23:45.

distance
noun

Distance is the length of the path between 2 points.

eg Measure the perpendicular distance between the 2 parallel lines.

distance–time graph
noun

A distance–time graph is a diagram showing how distance varies with time.

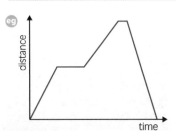

distribution
noun

A distribution is a collection of measurements or data.

eg Philip collected all the information about the cars he observed. He then drew up a frequency table to show the distribution of his groups.

distributive
adjective +√x−÷

An operation is distributive when a number multiplying the sum of 2 or more other numbers is equal to the sum of the products of the first number with each of the other numbers. This also applies when the numbers are subtracted instead of added.

eg Multiplication is distributive over addition and subtraction: $a(b + c) = ab + ac$, and $a(b − c) = ab − ac$.

divide
verb +√x−÷

To divide is to carry out the operation of division, or sharing a quantity into a number of equal parts.

eg Divide 64 by 16.

divisible
+√x−÷

One number that can be divided exactly by another number is said to be divisible by the second number.

eg 27 is divisible by 3.

➡ indivisible

division – draw

division [÷]

1 noun +√x‾÷

Division is an operation on numbers in which the number to be divided is shared equally into the stated number of parts.

eg Division is the opposite of multiplication.

2 noun

A division is one part of a scale.

eg Each division on the ruler represents a millimetre.

divisor +√x‾÷

A divisor is a number that divides another.

eg 3 is the divisor in the sum 27 ÷ 3.

dodeca- 1 2 3 4
prefix

Dodeca- is a prefix meaning 12.

dodecagon
noun

A dodecagon is a 12-sided polygon.

eg

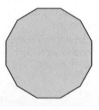

dodecahedron
noun

A dodecahedron is a 12-faced solid figure.

eg The faces of a regular dodecahedron are regular pentagons.

double +√x‾÷

1 adjective

Twice something is double.

eg When Tim worked at the weekend he was paid double wages.

2 verb

To double something is to multiply it by 2.

eg Think of a number, double it and subtract 6.

doubt
noun

If something is in doubt, it is uncertain.

eg When an event is in doubt, the probability is nearer to 0 than 1.

➡ uncertain

dozen 1 2 3 4
noun

A dozen is a group of 12.

eg The baker made a dozen gingerbread men.

draw

1 verb

To draw something out is to pull it out.

eg Draw out 4 balls from the bag and find the probability of picking 4 different colours.

2 verb

When you draw something you use a writing instrument to represent it.

eg Draw the graph in your class book.

Ee

earn 1 2 3 4
verb

To earn is to obtain payment for work or a service.

eg How much will Ben earn if he works 30 hours at £6.50 an hour?

➡ **salary, wage**

east
noun

East is one of the 4 main points of the compass, 90° clockwise or 270° anticlockwise of north.

eg

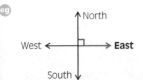

edge
noun

The edge of an object is where 2 or more faces meet.

eg

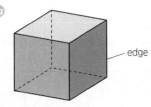

edge

either
1 conjunction

Either is used with 'or' when two alternatives are possible.

eg You may draw either a bar chart or a pie chart.

2 adjective

Either can also be used to mean both.

eg Measure the lengths of the curtains on either side of the window.

elder 1 2 3 4
adjective

The elder of two people is the one born first.

eg John was the elder twin, as he was born 5 minutes before James.

elevation
noun

The elevation of a building is a 2-dimensional vertical view seen from one side.

eg

eliminate 1 2 3 4
verb

To eliminate is to remove.

eg Eliminate the common factor from $\frac{12}{16}$ by cancelling.

ellipse
noun

An ellipse is an oval shape.

eg

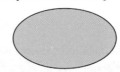

employ 1 2 3 4
verb

To employ someone is to pay them for work done.

eg Jan was employed by the company at a salary of £13,500 per year.

Employ comes from the Old French word emploier, *which means 'to work' or 'to be involved'.*

E

employee **1 2 3 4**
noun

Someone who does paid work for someone else is an employee.

eg The firm has 150 employees.

employer **1 2 3 4**
noun

Someone who employs other people is an employer.

eg She was the largest employer in the town.

enlargement
noun

An enlargement is a transformation of a figure by a given scale factor.

eg

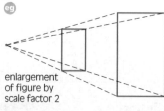

enlargement
of figure by
scale factor 2

➡ **scale factor**

enough
adjective

Enough means sufficient for requirements.

eg Will 4 cans of paint be enough to cover the walls?

enter
verb

To enter means to put data into a computer.

eg Ben entered the data onto the spreadsheet.

equally likely
adjective

When two events are equally likely, it means they have the same probability.

eg When a coin is tossed, a head and a tail are equally likely outcomes.

equals [=] **1 2 3 4**
verb

If one amount equals another, they are the same.

eg 150 pence equals £1.50.

equation $x = a^2$
noun

An equation is a mathematical statement showing things that are equal.

eg Solve the equation $5x = 25$ to show the value of the unknown.

equator
noun

The equator is the circle at 0° latitude that divides the Earth into 2 hemispheres.

eg

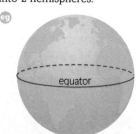

equator

equiangular
adjective

An equiangular polygon has all angles equal.

eg

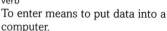

➡ **equilateral**

equidistant
adjective

A point is equidistant from 2 or more points, or lines, if it is the same distance from each of them.

eg A midpoint of a line is equidistant from both ends.

equilateral
adjective

Equilateral tells you that all sides are equal.

An equilateral polygon, known as a regular polygon, has equal sides, but not necessarily equal angles.

➡ **equiangular**

equivalent
1 2 3 4
adjective

If one or more quantities have the same value, but different forms, they are said to be equivalent.

1 kilometre is equivalent to 1000 metres.

equivalent fraction
noun
1 2 3 4

An equivalent fraction is a fraction with the same value as another.

$\frac{1}{2} = \frac{2}{4} = \frac{4}{8}$ are equivalent fractions.

error
$+\sqrt{x}\ ÷$
noun

An error is a mistake or fault in a process.

The calculation contained two minor errors.

estimate
$+\sqrt{x}\ ÷$
verb

To estimate is give an approximation of the actual value.

Estimate the answer to the following calculation.

estimate of mean
noun

An estimate of mean is used when analysing grouped frequency distributions, as the actual mean is not known. The midpoint of each class is used to estimate the mean.

Estimate of mean $(\Sigma fx ÷ \Sigma f)$ = (sum of products of frequency × midpoint for each class) ÷ (total frequency)

evaluate
? ? ? ?
verb

To evaluate is to assess the value of something.

Evaluate the VAT at 17.5% on a bill of £22.50.

even chance
➡ **chance**

even number
1 2 3 4
noun

When an even number is divided by 2, there is no remainder.

2, 4, 6, 8, 10… are all even numbers.

➡ **odd number**

event
noun

An event is an occurrence, happening or outcome.

The probability of a 2 being thrown with a dice is one of 6 possible events or outcomes.

➡ **occurrence**

evidence
? ? ? ?
noun

An indication of something happening or being the case is evidence of that event or situation.

The evidence that a line is straight is that the angles on the line add up to 180°.

exact
$+\sqrt{x}\ ÷$
adjective

An exact answer is the true value.

Give an exact answer, not a rough approximation.

examine
verb

To examine something is to inspect it closely.

Naomi examined the answer and discovered that she had missed out the decimal point.

E

A
B
C
D
E
F
G
H
I
J
K
L
M
N
O
P
Q
R
S
T
U
V
W
X
Y
Z

example
noun   **? ? ? ?**

An example is a specimen.

eg A trapezium is an example of a quadrilateral.

exception
noun

An exception is something that is excluded.

eg All prime numbers are odd numbers. The exception is 2, which is the only even prime number.

excess
adjective

An excess amount is surplus to requirements.

eg Water was poured from a 1 litre jug into a 0.75 litre container. The excess liquid measured 0.25 litres.

exchange
verb   **? ? ? ?**

To exchange is to swap.

eg Tim exchanged his 15 cm ruler for a 30 cm ruler.

exchange rate
noun   **$+\sqrt{x}-\div$**

The exchange rate is the rate at which the currency of one country is exchanged for that of another.

eg How many dollars will Hamid receive for £150 at an exchange rate of £1 = $1.40?

➡ **foreign exchange**

exercise
noun   **1 2 3 4**

An exercise is an activity used to practise a procedure.

eg Exercise 4 consists of 20 questions.

exhaustive
adjective

If something is exhaustive, it covers every possibility.

eg The investigation was exhaustive.

expand
verb   **$x = a^2$**

To expand, or open, brackets means to multiply them out.

eg Expand the brackets and simplify the expression $3(x - 1) - 2(4 + x)$.

➡ **multiply out**

expand the product
phrase   **$x = a^2$**

To expand the product of 2 linear expressions is to multiply each term of one by each term of the other.

eg Expand the product of $(x + 2)(x - 3)$ and simplify.

expensive
adjective   **1 2 3 4**

If something is expensive it costs a lot of money.

eg The coat was far too expensive even after a sale discount of 25%.

➡ **cheap, dear**

experiment
noun

An experiment is a test or trial.

eg Their experiment was tossing two coins 100 times to check the probability of getting 2 heads.

experimental probability
noun

When the probability is found by the event being carried out many times, it is called experimental probability.

eg Experimental probability = (number of successful outcomes) ÷ (total number of events)

explain
verb   **? ? ? ?**

To explain is to define what is happening.

eg Explain how you have worked out the angles.

explore
verb   **? ? ? ?**

To explore is to investigate carefully.

Toss 3 coins and explore the different combinations of heads and tails.

exponent
➡ index

express
verb   **1 2 3 4**

To express something is to state it in certain terms.

Express 0.000785 in standard form.

expression
noun $x = a^2$

An expression is an algebraic statement having letters and numbers.

$6x^2 + 2xy - y^2$ is an expression, not an equation, as it does not have 2 statements which are equal.

➡ equation

extend
verb

To extend is to lengthen or continue.

Draw a triangle and extend the base line.

exterior
noun

The exterior is the outside.

The surface area of a cylinder is the area of its exterior.

➡ interior

exterior angle
noun

An exterior angle is the angle between the extended side of a polygon and the adjacent side.

exterior angle

➡ interior angle

external
adjective

External means to do with the exterior or outside.

The sum of the external angles of a polygon is 360°.

➡ internal

extract
verb

To extract is to find particular examples, data or information from a source.

Extract the multiples of 3 from the following numbers: 27, 38, 57, 62.

extrapolate
verb

To extrapolate is to predict results beyond the extent of the given values.

Adam used his heating bills for the past 2 years to extrapolate his bill for the next year.

➡ interpolate

extreme
noun

An extreme is an end value, either highest or lowest.

The temperature dropped to an extreme of −25°C.

Ff

face

1 noun

The flat surface, or side, of a solid shape is called a face.

eg A cuboid has 6 faces.

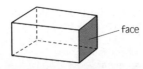
— face

2 verb

To face is to be opposite something.

eg The hypotenuse faces the right angle in a right-angled triangle.

factor

noun

1234

A factor is a whole number which exactly divides another whole number.

eg 3 and 5 are factors of 15.

factorise

1 verb

1234

To factorise means to break up, or separate, into factors.

eg 24 may be factorised into $2 \times 3 \times 4$ or 4×6 or 3×8.

2 verb

$x = a^2$

The term factorise is used with the same meaning in algebra.

eg $x^2 + 5x + 6$ may be factorised into $(x + 2)(x + 3)$.

Fahrenheit (*fa-ren-hite*)

adjective

Fahrenheit refers to the Fahrenheit scale, used for measuring temperature.

eg On the Fahrenheit scale the boiling point of water is 212°F and the freezing point of water is 32°F.

This temperature scale is named after G. D. Fahrenheit, a German physicist at the beginning of the 18th century. To change temperature from Fahrenheit to Celsius:

$°C = \frac{5}{9}(°F - 32)$.

➡ **Celsius**

fair

adjective

Fair describes an item or event which is not biased.

eg Heads and tails were equally likely when the fair coin was tossed.

➡ **unfair**

false

adjective

? ? ? ?

False means not true.

eg The temperature today is 32°C. True or false?

fare

noun

1234

When you travel on a form of transport, you must pay a fare.

eg The normal train fare was £36.50.

fault

noun

$+\sqrt{x}-\div$

A fault is a mistake or error in a process.

eg There was a fault in the working which produced an error in the final answer.

favourite

adjective

A favourite object is the chosen one.

eg Year 7 conducted a survey to find people's favourite TV programme.

Fibonacci sequence
noun **1234**

A Fibonacci sequence is one where each number is found by adding the 2 previous numbers.

🔹 1, 1, 2, 3, 5, 8, 13, 21... is a Fibonacci sequence.

🔹 Leonardo Fibonacci was an Italian mathematician who lived at the beginning of the 13th century.

fifty-fifty
adjective, adverb

When there is an equal chance of an event happening or not happening, it is said to be fifty-fifty.

🔹 There's a fifty-fifty chance of the coin turning up heads.

figure
1 noun

A figure is a shape or object.

🔹 A plane figure with 3 sides is a triangle.

2 noun **1234**

A figure can also mean a numerical symbol.

🔹 In the number 2314, the figure 3 represents 300.

finance
1234
verb

To finance a project is to provide money for it.

🔹 The building society financed the purchase of their house.

finite (*fi-nite*)
1234
adjective

Finite means limited.

🔹 The sequence was continued to a finite end.

➡ **infinite**

first [1st]
1234
adjective

The first is the earliest in order.

🔹 The first prime number is 2.

first/second difference
noun $x = a^2$

Terms in sequences that do not have a constant difference are generated using first and second differences.

🔹 You need to use first difference and second difference when generating terms in the sequence 6, 15, 28, 45, 66...

6   15   28   45   66
  9   13   17   21   (first difference)
    4   4   4   (second difference)

➡ **difference pattern**

flat
adjective

Something that is flat is level or even.

🔹 Stand the model on a flat surface.

flexible
adjective

A flexible material is supple or elastic.

🔹 The cube was modelled using flexible card.

flow chart
noun $x = a^2$

A flow chart is a diagram illustrating the sequence of operations in a procedure. It is sometimes called a function machine.

🔹

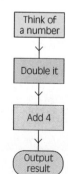

foot – frequency chart

foot (plural: feet) [ft]
noun

A foot is an imperial unit of length equal to 12 inches.

eg There are 3 feet in 1 yard. 3 ft ≈ 1 m

A foot was based on the length of the average human foot.

force
noun

A force is a power felt when there is pressure being exerted.

eg Gravity is a force we all feel on Earth.

forecast
verb

To forecast is to predict an occurrence.

eg After the class tests, the teacher could forecast his pupils' results in the Key Stage 3 tests.

foreign exchange $+\sqrt{x}\div$
noun

Foreign exchange is the system by which the currency of one country is exchanged for that of another.

eg The foreign exchange rates gave £1 = 0.6122 euros.

➡ exchange rate

format
noun

The format of an object is its size and shape according to given requirements.

eg Her investigation was presented in the format of a book.

formula (plural: formulae or formulas) $x = a^2$
noun

A formula is an equation used to find quantities when given certain values.

eg The formula for converting °F to °C is: °C = $\frac{5}{9}$(°F − 32).

fortnight
noun

A fortnight is 2 weeks or 14 days.

eg There is a fortnight before the Key Stage 3 examination.

forward
adverb

If you move forward you go ahead.

eg The boy moved forward in the queue.

foundation 1 2 3 4
noun

The foundation of something is its base or groundwork.

eg Learn your tables so that you have a good foundation for multiplication and division sums.

fraction 1 2 3 4
noun

A fraction is a part of the whole.

eg $\frac{5}{6}$, $\frac{11}{3}$ and $2\frac{3}{4}$ are all fractions.

➡ common fraction, decimal fraction, improper fraction, mixed fraction

frequency
noun

The number of times that something happens is called the frequency.

eg The frequency of trains is 2 every hour.

frequency chart
noun

A frequency chart, or frequency diagram, is an illustration of the frequency of a given event.

eg

frequency

frequency diagram – further

frequency diagram
➡ frequency chart

frequency polygon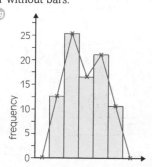
noun

A frequency polygon joins the midpoints of data groups or classes in a continuous distribution. It can be drawn with or without bars.

eg

frequency table
noun

A frequency table is an arrangement of data in columns.

eg

mark	frequency
1–10	6
11–20	10
21–30	16
31–40	20
41–50	22
51–60	20
61–70	6

frequent
adjective

Something that is frequent happens often.

eg The most frequent is the mode.

function $x = a^2$
noun

A function is a relationship between variables. The first variable must depend on the other variable.

eg y is a function of x in the equation $y = x^2 + 2x - 1$.

function machine
➡ flow chart

further $1\,2\,3\,4$
adjective/adverb

Further means more or in addition.

eg Write down further terms in the sequence.

G g

gallon [gal]
noun

A gallon is an imperial unit of capacity.

eg There are 8 pints in 1 gallon.

generalise ? ? ? ?
verb

To generalise is to form a general statement or rule.

eg Generalise the results of the experiment and test by checking particular cases.

general statement $x = a^2$
noun

A general statement says what usually applies in all relevant cases.

eg Make a general statement based on the evidence produced.

general term $x = a^2$
noun

A general term is a rule that applies to every term in a sequence.

eg The general term, or nth term, of the sequence 0, 2, 6, 12, 20… is $n(n-1)$.

generate $x = a^2$
verb

To generate is to create.

eg Generate a sequence starting with 3 and adding 5 each time.

geo-board
noun

A geo-board, originally called a geometry board, is a flat board with regular rows and columns of nails or pegs. Elastic bands are stretched around these pegs to create polygons.

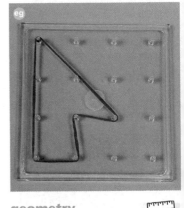

geometry
noun

The study of shape and space is called geometry.

eg In geometry we study angles, triangles, circles and polygons.

giga- [G]
prefix

Giga- is a prefix meaning 1 000 000 000.

good chance
➡ chance

grade 1 2 3 4

1 noun

A grade is a level attained.

eg Harry achieved a grade of B+ in his exam.

2 verb

To grade objects is to put them into order.

eg The questions were graded into Levels 3–5, 4–6, 5–7 and 6–8.

gradient $x = a^2$

noun

A gradient is a measure of the steepness of a slope.

Gradient = (vertical distance) ÷ (horizontal distance)

eg A graph that goes up as it moves to the right has a positive (+) gradient; a graph that goes down as it moves to the right has a negative (−) gradient.

gradual

adjective

A gradual progression is one that proceeds at a regular, often slow, rate.

eg The plotted points were joined to form a gradual curve.

gram [g]

noun

A gram is a unit of weight or mass in the metric system.

eg There are 1000 grams in 1 kilogram.

graph $x = a^2$

noun

A graph is a way of illustrating a relationship between variables.

eg A journey can be illustrated by plotting distance against time and producing a travel graph.

➡ **bar chart, cubic curve, parabola, pie chart**

graph paper $x = a^2$

noun

Graph paper has measured ruled lines, usually in 1 cm squares, to enable graphs to be drawn easily.

eg

gravity

noun

Gravity is the force that the Earth exerts on any object.

eg If you drop something, gravity pulls it towards the ground.

H The English scientist Isaac Newton is famous for developing a number of theories about gravity. He found his inspiration after watching an apple fall to the ground in his mother's orchard.

greater than [>] 1234

phrase

If one quantity is larger than another, the first is greater than the second.

eg Is $\frac{3}{4}$ greater than $\frac{5}{6}$?

➡ **less than**

greater than or equal to [≥] 1234

phrase

If one quantity can be larger than or equal to another, the first is greater than or equal to the second.

eg $(6 + y)$ is greater than or equal to 7.

➡ **less than or equal to**

greatest value 1234

noun

The greatest value, or maximum value, is the largest amount or quantity.

eg What is the greatest value number that can be created from the digits 1, 2, 3, 4?

➡ **least value**

grid

$x = a^2$

noun

A grid is a network formed by parallel vertical and horizontal lines used for locating points on maps or charts.

eg

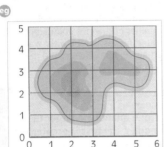

gross

1 2 3 4

1 adjective

A gross amount is the total without deductions.

eg Helen's gross pay was £12,500 before income tax was deducted.

➡ net

2 noun

A gross is 12 dozen (144).

eg A gross is $12 \times 12 = 12^2 = 144$.

grouped data

noun

Grouped data is information organised into groups.

eg The grouped data had intervals of 5 cm.

➡ class

A
B
C
D
E
F

G

H
I
J
K
L
M
N
O
P
Q
R
S
T
U
V
W
X
Y
Z

Hh

half (plural: halves)
(*harf*)   **1 2 3 4**

1 noun

A half is either of 2 equal parts into which a whole is divided.

eg A diameter divides a circle into 2 halves.

2 noun

Half also means the fraction $\frac{1}{2}$.

eg What is half of 76?

halve (*harv*) $+\sqrt{x}\,\overline{}\div$
verb

To halve something is to divide it into 2 equal parts.

eg Halve the rectangle to create 2 right-angled triangles.

hectare [**ha**] (*hek-tair*)
noun

A hectare is a unit of area measurement in the metric system: 1 hectare = 10 000 square metres.

eg A square of side 100 metres covers an area of 1 hectare.

hecto-
prefix

Hecto- is a prefix meaning 100.

height (*hite*)
noun

The height of something is how tall it is.

eg The height of the door was 200 cm, its width 80 cm.

helix (*hee-lix*)
noun

A helix is a spiral curve like one drawn on the surface of a cone or cylinder.

H Helix comes from the Greek word *helissein*, meaning 'to twist'.

hemisphere
noun

A hemisphere is half of a sphere.

eg

heptagon
noun

A heptagon is a 7-sided polygon.

eg

hexagon
noun

A hexagon is a 6-sided polygon.

eg

A B C D E F G **H** I J K L M N O P Q R S T U V W X Y Z

high

1 adjective

How high an object is tells you how far it is off the ground.

eg The shelf was too high to reach.

2 adjective

How high an object is tells you how tall it is from top to bottom.

eg The wall was 1.3 metres high.

3 adjective **1234**

High also refers to the value of numbers.

eg When $x = 3$ was substituted in the formula, the result was too high.

highest common factor [HCF] **1234**

noun

The highest common factor is the highest factor shared by 2 or more numbers.

eg What is the highest common factor of 15 and 60?

Hindu-Arabic number system **1234**

noun

The Hindu-Arabic number system is a number system which uses the digits 0–9.

eg The Hindu-Arabic number system uses base 10.

hire **1234**

verb

When you hire something, you make a payment for its use.

eg How much will it cost to hire a car for 4 days at a rate of £33 per day?

hire purchase **1234**

noun

When you buy something on hire purchase, you pay a deposit and the balance in instalments at a given rate of interest.

eg A TV costing £500 is bought on hire purchase. If 10% interest is charged, and a deposit of £50 paid, how much has to be paid in each of 10 monthly instalments?

horizontal

1 noun

The horizontal is a straight level line perpendicular to the vertical.

eg An angle of depression is measured from the horizontal down to the line of sight of an object.

2 adjective

Horizontal means parallel to the horizon.

eg Draw a horizontal line 3 cm long.

hour

noun

An hour is a measurement of time consisting of 60 minutes.

eg There are 24 hours in a day.

➡ **time**

hundredth [100th] **1234**

1 adjective

Hundredth is the ordinal number of a hundred.

eg On its hundredth throw the coin turned up tails.

2 noun

A hundredth is the fraction $\frac{1}{100}$.

eg A penny is a hundredth of a pound.

hypotenuse (*hie-**pot**-a-news*)

noun

The hypotenuse is the side opposite the right angle in a right-angled triangle.

eg

➡ **Pythagoras' theorem**

hypothesis (*hie-**poth**-i-siss*)

noun **????**

A hypothesis is a theory that is tested, by investigation, to see if it is true.

eg How could you test the hypothesis that girls and boys like the same music

I i

1 2 3 4

I is the symbol which stands for 1 in the Roman number system.

eg C + VI + IV + III = CXIII

➡ **Roman numerals**

icosahedron

(*ike-u-sa-**heed**-ron*)

noun

An icosahedron is a 20-faced solid figure.

eg The faces of a regular icosahedron are equilateral triangles.

ideal

adjective

An ideal solution is the best possible or the most perfect solution.

eg Hexagons produce an ideal tessellation.

identical

adjective

Identical means the same.

eg Congruent figures have identical dimensions.

identically equal to [≡]

phrase

Identically equal to means exactly the same.

eg Congruent figures are identically equal to one another.

identify

verb

To identify is to recognise or name something or somebody.

eg Identify the square numbers in the following: 15, 25, 30, 36, 45, 49.

identity $x = a^2$

noun

If an equation is true for all values of its variables it is called an identity.

eg $a(b + c) \equiv ab + bc$ is an identity.

identity function $x = a^2$

noun

The relationship which maps any element x of set X into itself is an identity function.

eg The identity function $x \rightarrow x$ maps a number on to itself, so the number remains unchanged.

➡ **function**

illustration

noun

An illustration shows the appearance of something or gives a picture of an event.

eg A pie chart is an illustration of data collected.

image

noun

When an object is transformed by reflection the result is an image.

eg

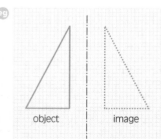

object | image

45

imperial unit
noun

Imperial units are units of weight and measurement, which have been more or less superseded by metric units.

eg Feet, yards and miles are imperial units of length.

➔ **metric unit**

impossible
adjective

If an event is impossible, it cannot happen.

eg The probability of an impossible event = 0.

➔ **probability**

improper fraction 1234
noun

An improper fraction is a fraction which has a numerator greater than the denominator.

eg $\frac{9}{4}$ is an improper fraction, which could be expressed as the mixed number $2\frac{1}{4}$.

➔ **denominator, numerator, proper fraction**

inch [in]
noun

An inch is a unit of length in the imperial system.

eg 12 inches = 1 foot; 1 in = 2.54 cm

🏴 An inch was originally defined by Edward II. It was the length of 3 barley grains end to end.

incline
noun

An incline is a slope.

eg The gradient of the incline of a ladder leaning against a wall is calculated as follows: the height reached by the ladder ÷ the distance of the foot of the ladder from the wall.

➔ **angle of inclination, gradient**

included
adjective

Included items are contained within certain limits.

eg Construct a triangle with sides 5 cm, 10 cm and an included angle of 70°.

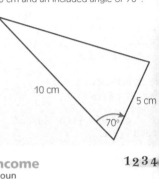

income 1234
noun

Income is the pay earned for work done.

eg Gill had a net income of £10,500 after deductions.

➔ **salary, wage**

increase +√x−÷
1 noun

An increase is an enlargement or addition.

eg An increase of 3.5 cm makes a total length of 9 cm.

2 verb

To increase is to grow or get bigger, or to make something bigger.

eg He increased the amount from £30: to £384.

➔ **decrease**

independent
adjective

Something is independent if it doesn't depend on anything else

eg A and B are independent events if they can happen together but do not depend on each other.

➔ **'and' rule**

index (plural: indices) $\mathbf{1}\underset{\sim}{2}\mathbf{3}\,\mathbf{4}$
noun

The index is the small digit to the top right of a number which tells you the number of times a number is multiplied by itself. Other words used for index are power and exponent.

(eg) $5^4 = 5 \times 5 \times 5 \times 5$. The index is 4.

index law $\mathbf{1}\underset{\sim}{2}\mathbf{3}\,\mathbf{4}$

An index law is a rule for manipulating index numbers when powers are multiplied or divided.

(eg) Certain index laws must be followed.

(1) When multiplying powers of the same term, add indices, e.g.
$a^3 \times a^2 = a^{(3+2)} = a^5$;
(2) When dividing powers of the same term, subtract indices, e.g.
$a^5 \div a^2 = a^{(5-2)} = a^3$;
(3) When raising a power of a term to another power, multiply indices, e.g. $(a^5)^2 = a^{10}$.

index notation $\mathbf{1}\underset{\sim}{2}\mathbf{3}\,\mathbf{4}$

Index notation is a shorter way of recording different products of numbers.

(eg) Certain index notations give specific results.

index < 0   (negative) $a^{-2} = \frac{1}{a^2}$
index = 0   (always = 1) $a^0 = 1$
index > 0   (positive) $a^3 = a \times a \times a$
0 < index < 1 (fraction) $a^{\frac{1}{n}} = \sqrt[n]{a}$
➡ **index**

indicate $\mathbf{1}\underset{\sim}{2}\mathbf{3}\,\mathbf{4}$
verb

To indicate is to point out.

(eg) When you have finished your scale drawing, label it to indicate the different measurements.

individual $\mathbf{?}\,\mathbf{?}\,\mathbf{?}\,\mathbf{?}$
noun

An individual is one person or object.

(eg) A protractor was given to each individual.

individually $x = a^2$
adverb

Individually means on its own.

(eg) Look at the algebraic expression and consider each term individually.

indivisible $\mathbf{1}\underset{\sim}{2}\mathbf{3}\,\mathbf{4}$
adjective

A whole number that cannot be divided exactly by another number is indivisible.

(eg) An odd number is indivisible by 2.
➡ **divisible**

inequality $[\neq < >]$
noun $x = a^2$

An inequality is a statement showing two quantities that are not equal.

(eg) $x \neq y$ and $x < y$ are inequalities.
➡ **equation, greater than, less than, region**

infinite (**in**-fin-it) $x = a^2$
adjective

Infinite means without limit, or having an unlimited number of digits or terms.

(eg) $\frac{1}{3}$, when converted to a decimal, has an infinite number of decimal places (0.33333…).
➡ **finite**

infinity $[\infty]$ $\mathbf{1}\underset{\sim}{2}\mathbf{3}\,\mathbf{4}$
noun

Infinity is a quantity larger than any known quantity.

(eg) A number increasing without limit tends to infinity.

information

noun

Information is facts or data.

eg The information about the heights of the seedlings was stored in a database.

➡ **data**

input

noun

Input is the information put into a task or into a computer.

eg Kirsty made sure she saved her input onto a disk.

inscribe

verb

If a figure is inscribed inside another figure it means that they touch at several points but do not intersect.

eg

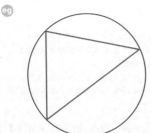

instruction

? ? ? ?

noun

An instruction is a command to obey certain rules.

eg Follow the instructions to complete the flow chart.

insurance

1 2 3 4

noun

An insurance company will protect your property or health against harmful events, if you pay them an amount of money called a premium.

eg After the house was flooded, the owner sent estimates for repair to the insurance company.

integer

1 2 3 4

noun

An integer is any whole number, positive or negative, including zero.

eg Which integers satisfy the inequality $-3 \leq a < 3$?

intercept

$x = a^2$

noun

An intercept is the point where a line crosses an axis on a graph.

eg

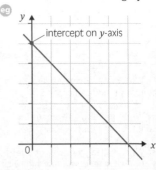

intercept on y-axis

interest

1 2 3 4

1 noun

Interest is the amount of money paid to you by a bank or building society when you invest savings with them.

eg There are two types of interest on savings: simple and compound interest

2 noun

Interest is also charged on money borrowed from a bank or building society.

eg What rate of interest are you paying on your mortgage?

All interest is calculated at a rate of interest based on Bank of England rates. It can be calculated monthly, but it is usual to calculate it yearly (annually). The original amount of investment or loan is called the principal.

➡ **compound interest, deposit, principal, simple interest**

interior

noun

The interior is the inside.

eg Girish measured the length of the interior of the box.

➡ **exterior**

interior angle

1 noun

An interior angle is the angle between one side of a polygon and the adjacent side.

eg

interior angle

➡ **exterior angle**

2 noun

Supplementary angles on a transversal between 2 parallel lines are called interior angles.

eg

$$\hat{a} + \hat{b} = 180°$$

➡ **supplementary angle**

internal

adjective

Internal means to do with the interior or inside.

eg The sum of the internal angles of a polygon is 360°.

➡ **external**

interpolate

verb

To interpolate is to predict other results within given values.

eg The students' shoe sizes were plotted against heights in a scatter graph and a line of best fit was drawn. This enabled Tony to interpolate what shoe sizes would be for other pupils of different heights.

➡ **extrapolate**

interpret

verb

To interpret is to explain.

eg Use the key to interpret the pictogram.

interquartile range

noun

The interquartile range is the difference between the lower quartile and the upper quartile when interpreting a cumulative frequency graph.

eg

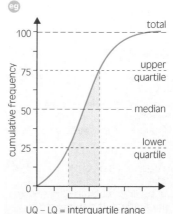

UQ – LQ = interquartile range

➡ **cumulative frequency graph, quartile**

interrogate

verb

To interrogate a database is to explore its contents.

eg Miriam needed to interrogate the database to find out how many items were in each group.

intersect

verb

To intersect is to cross at a point.

eg When 2 lines intersect they form 4 angles at the point of intersection.

interval

1 noun

An interval is a period of time.

eg There is an hour's interval between 13.00 and 14.00.

2 noun

An interval is the difference between two or more numbers.

eg There is an interval of 5 between the numbers 4 and 9.

invalid $x = a^2$

adjective

If something is invalid, it does not obey a given rule or condition.

eg The equation was invalid for negative numbers. It was only satisfied by positive numbers.

inverse $+\sqrt{x} - \div$

noun

The inverse is the reverse or opposite.

eg Addition is the inverse of subtraction.

inverse function $x = a^2$

noun

An inverse function exists when one function is the inverse of another function.

eg Dividing by 5 is the inverse function of multiplying by 5.

➡ function

inverse mapping $x = a^2$

noun

An inverse mapping reverses the direction of the original mapping.

eg Inverse mapping will only work if there is a one-to-one correspondence.

➡ mapping

inverse proportion 1 2 3 4

noun

Inverse proportion is a relationship between 2 variables such that the value of 1 increases as the other decreases.

eg The distance a car travels varies in inverse proportion to the amount of petrol in its tank.

➡ direct proportion

investigate ? ? ? ?

verb

To investigate something is to try to find out the facts about it.

eg Design a questionnaire to help you investigate the shopping habits in your town.

irregular

adjective

Irregular objects do not follow a given pattern or format.

eg A scalene triangle is irregular as it has 3 different sides and angles.

➡ regular

isometric drawing

noun

An isometric drawing is a 3D representation of an object in which the 3 axes are equally inclined and all lines are drawn to a given scale.

eg

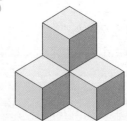

isosceles trapezium

(*ice-**oss**-ill-eez trap-**ee**-zee-um*)
noun

An isosceles trapezium
is a quadrilateral with 1 pair of
parallel sides, 1 pair of equal,
non-parallel sides, equal
diagonals and 2 pairs of equal
angles.

eg

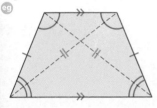

➡ trapezium

isosceles triangle

noun

An isosceles triangle has 2 equal
sides with the opposite angles
also equal.

eg

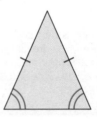

Isosceles comes from two Greek
words, *isos* meaning equal and
skelos meaning leg.

item

noun

An item is one of a number of
things.

eg How do you calculate the mean of a
number of items?

justify
???

verb

When you justify a decision or conclusion, you provide good reasons to support it.

eg You must give evidence to justify your conclusions.

Kk

key
1 noun
A key is a button on a keyboard.

2 noun
A key to a chart tells you what each symbol means.

eg Use the key to interpret the pictogram.

kilo-
prefix
Kilo- is a prefix meaning 1000.

kilobyte [KB] (*kil-u-bite*)
noun
A kilobyte is 1024 (2^{10}) bytes.

eg Computer memory size is measured in kilobytes or megabytes.
➡ byte, megabyte

kilogram [kg]
noun
A kilogram is a measurement of weight or mass equal to 1000 grams.

eg How many kilograms are in 2500 grams?

kilolitre [kl]
noun
A kilolitre is a measurement of capacity equal to 1000 litres.

eg How many kilolitres are there in 3450 litres?

kilometre [km]
noun
A kilometre is a measurement of length equal to 1000 metres.

eg 1 kilometre is approximately $\frac{5}{8}$ mile.

kite
noun
A kite is a quadrilateral with 2 pairs of equal adjacent sides, 1 pair of equal opposite angles and diagonals intersecting at right angles.

eg

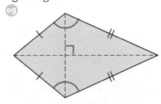

knot (*not*)
noun
A knot is a measure of speed equalling 1 nautical mile per hour or 1.15 miles per hour.

eg The ship moved at a speed of 30 knots.

i The nautical measure of speed was measured by a rope or log line, divided by knots at equal distances ($\frac{1}{120}$ of a geographical mile). The number of knots travelled by a ship in $\frac{1}{2}$ a minute corresponded to the number of nautical miles travelled per hour.

L1

L 1234

L is the symbol which stands for 50 in the Roman number system.

eg LV − VI + XIV = LXIII

➡ **Roman numerals**

label

verb

To label is to add a label or tag giving information.

eg Sue labelled each angle and length in her diagram.

Latin square 1234

noun

In a Latin square, the numbers appear only once in each row and each column.

eg

8	3	4
1	5	9
6	7	2

latitude

noun

Latitude is a line parallel to the equator.

eg

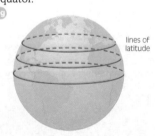

lines of latitude

➡ **longitude**

LCD

➡ **lowest common denominator**

LCM

➡ **lowest common multiple**

leap year

noun

A leap year is a calendar year of 366 days in which February has 29 days instead of 28. A leap year occurs every fourth year.

eg The year 2002 is not a leap year as it is not divisible by 4.

➡ **year**

least 1234

adjective

The least amount is the smallest quantity.

eg What is the least amount needed to add to 17 to make a square number?

least common multiple

➡ **lowest common multiple**

least significant digit

noun 1234

The digit that is not taken into account when correcting to a given number of decimal places or significant figures is called the least significant digit.

eg When you compare the size of 2.65<u>4</u> with 2.68, the underlined 4 is the least significant digit. The second decimal place in each number tells you that 2.68 > 2.65, so the 4 is not needed for the comparison.

➡ **most significant digit**

least value 1234

noun

The least value, or minimum value, is the smallest amount or quantity.

eg The least value of a square number is 1 as it can never be negative.

lend

1234

verb

To lend is to allow someone to use your money or property.

eg When the building society lent Holly money to help her buy a car, she had to pay interest of 7.5%.

➡ **borrow**

length

noun

A length is a distance between 2 points.

eg Length is measured in millimetres, centimetres, metres and kilometres in the metric system.

➡ **breadth, width**

less than [<]

1234

phrase

If one quantity is smaller than another, the first is less than the second.

eg Is $\frac{2}{3}$ less than $\frac{5}{6}$?

➡ **greater than**

less than or equal to [≤]

phrase **1234**

If one quantity is smaller than or the same as another, the first is less than or equal to the second.

eg $6y$ is less than or equal to 5.

➡ **greater than or equal to**

level

1 noun

A level is a particular grade on a scale.

eg The level of the mercury in the thermometer reached 15°C.

2 adjective

A level surface is one that is flat and horizontal.

eg The carpenter checked that the shelf was level.

➡ **flat, horizontal**

likelihood

noun

The likelihood of an event occurring is how probable it is.

eg The likelihood of getting a 6 when throwing a dice is $\frac{1}{6}$.

likely

adjective

If an event is likely it is probable.

eg A likely event will have a probability between 0.5 and 1.

➡ **probability**

like terms

$x = a^2$

noun

When terms in an algebraic expression are alike they can be gathered together; this is called collecting like terms.

eg Simplify the expression by collecting like terms: $7a^2 + 2b - 3a^2 - 5b + a = 4a^2 + a - 3b$.

limit

1234

noun

A limit is a boundary, or the ultimate quantity or extent of something.

eg The limit of the sequence of fractions $\frac{1}{2}$, $\frac{2}{3}$, $\frac{3}{4}$, $\frac{4}{5}$... is 1.

line

noun

A line is a straight or curved length with no width.

eg A line was drawn from the vertex of the triangle to the midpoint of the opposite side.

linear equation

$x = a^2$

noun

A linear equation is one having no variable above the power 1.

eg $y = 5x + 2$ is a linear equation.

linear expression

$x = a^2$

noun

A linear expression is one having no variable above the power 1.

eg $4a + b + 3c - 7$ is a linear expression.

linear function $x = a^2$
noun

A linear function, or linear relationship, has no variable above power 1 and expresses one unknown in terms of another unknown.

eg $2y + 3 = 7x$ is a linear function.

linear graph $x = a^2$
noun

A linear graph is a graph of a linear function, where all plotted points lie on a straight line.

eg

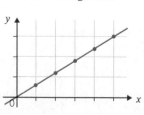

linear relationship
➡ linear function

linear sequence $x = a^2$
noun

A linear sequence is generated by increasing or decreasing by the same amount.

eg 4, 8, 12, 16… is a linear sequence increasing by 4 each time.

line graph $x = a^2$
noun

A line graph is a graph where all plotted points are joined by straight lines.

eg

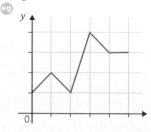

line of best fit
phrase

The line of best fit, which is drawn approximately in the middle of the points of a scatter diagram, enables you to estimate values not given in the original information.

eg

line of symmetry
noun

A line of symmetry is a line about which a shape is symmetrical.

eg

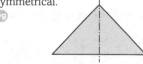

➡ mirror line, reflection

line segment
noun

A piece of a straight line is a line segment.

eg

line segment

line symmetry
noun

Line symmetry, or reflection symmetry, is the symmetry of a plane, or 2D, shape giving 2 equal halves.

eg The letter 'W' has line symmetry.

litre [1]
noun

A litre is a metric unit of capacity

eg There are 1000 millilitres in 1 litre.

loan $1\,2\,3\,4$
noun

A loan is property or money borrowed from somebody else or lent to somebody else.

eg A loan from a building society, to buy a property, is called a mortgage.

➡ **borrow, interest, lend, mortgage**

locus (plural: loci)
noun

A locus is the path of a point that moves according to a given rule.

eg The locus of a point moving at a fixed distance (r) from a given point (A) is a circle with centre A and radius r.

long division $+\sqrt{x}-\div$
noun

Long division is division of a number by another ≥ 10.

eg $1350 \div 18$ is an example of long division.

longitude
noun

Longitude is the distance in degrees east or west of the prime meridian at $0°$.

eg Lines of longitude are drawn at right angles to lines of latitude.

➡ **latitude, meridian**

long multiplication
noun $+\sqrt{x}-\div$

Long multiplication is multiplication of a number by another ≥ 10.

eg 273×13 is an example of long multiplication.

loss $1\,2\,3\,4$
noun

In business, to make a loss is to not make a profit and to lose money on a deal.

eg A shopkeeper bought 4 boxes, each containing 50 pens, for £60. He sold only 3 boxes for 35p a pen. Did he make a profit or a loss?

➡ **profit**

lower bound $x = a^2$
noun

The lower bound is the bottom limit.

eg The lower bound of all numbers satisfying the inequality
$1.75 \leq n < 3.5$ is 1.75.

➡ **upper bound**

lower quartile
➡ **quartile**

lowest common denominator [**LCD**] $1\,2\,3\,4$
noun

The lowest common denominator is the lowest common multiple of all the denominators in a set of fractions.

eg In the following sum, 56 is the lowest common multiple of 7 and 8 and therefore is the lowest common denominator:

$$\frac{3}{8} + \frac{4}{7} = \frac{(7 \times 3)}{56} + \frac{(8 \times 4)}{56}$$

$$= \frac{21 + 32}{56} = \frac{53}{56}$$

lowest common multiple [**LCM**] $1\,2\,3\,4$
noun

The lowest common multiple is the lowest number which is a multiple of 2 or more numbers.

eg The lowest common multiple of 18 and 60 is 180.

lowest terms $1\,2\,3\,4$
noun

If a fraction is in its lowest terms, it has been cancelled until the only common factor of both numerator and denominator is 1.

eg Put $\frac{48}{160}$ in its lowest terms.

➡ **cancel, simplest form**

lozenge
➡ **rhombus**

Mm

M   1234

M is the symbol which stands for 1000 in the Roman number system.

eg MCMLXXXVI = 1986

➡ **Roman numerals**

magic square   1234
noun

A magic square is an arrangement of numbers in the form of a square, in which each column and each row have the same sum.

eg

6	6	3
2	5	8
7	4	4

magnitude   1234
noun

The magnitude of something is its size.

eg The magnitude of the 3rd angle of an isosceles triangle with base angles of 35° is 110°.

main   ????
adjective

The main item is the most important or most frequent item.

eg The pie chart shows that the main TV viewing for Year 8 is from 19.00 to 20.30.

majority   1234
noun

A majority of a group is more than half the items or persons in that group.

eg In the election the winning political party was the one with a majority of the votes.

➡ **minority**

major sector
noun

The major sector is the larger section of the circle between 2 radii and an arc.

eg

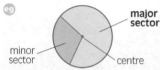

➡ **minor sector, sector**

major segment
noun

The major segment is the larger section of the circle between a chord and an arc.

eg

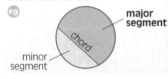

➡ **minor segment, segment**

map
noun

A map is a diagram giving geographical details.

eg

mapping $x = a^2$
noun

A mapping is changing something by following a given rule.

eg The mappings 2→8, 3→27, 4→64 obey the rule 'cube number'.

mass
noun

The mass of an object is the amount of material contained in it.

(eg) Milligrams, grams, kilograms and tonnes are units of mass.

➡ **weight**

mathematics [**maths**]
noun

Mathematics is the study of number and space.

(eg) Mathematics consists of different topics like algebra and geometry.

maximum point
noun

$x = a^2$

The maximum point on a curve is the top of that curve.

(eg)

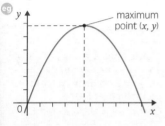

maximum point (x, y)

➡ **minimum point**

maximum value
➡ **greatest value**

mean
noun

The mean, or arithmetic mean, is an average value found by dividing the sum of a set of quantities by the number of quantities.

(eg) The mean of 11, 14, 17 and 18 is found like this:

$$\frac{11 + 14 + 17 + 18}{4} = \frac{60}{4} = 15$$

➡ **average, median, mode**

meaning
noun

The explanation of something is its meaning.

(eg) The meaning of x^2 is 'multiply x by itself'.

measure
verb

To measure something is to find the size, quantity or degree of it.

(eg) Yasmin measured the room and found that is was 3.5 m by 5 m.

measurement
noun

A measurement is an amount or size discovered by measuring.

(eg) The measurement of the diameter was found to be 15.5 cm.

median
noun

The median is the middle item in an ascending sequence of items.

(eg) What is the median of 14, 13, 13.5, 11, 16.5, 17, 15 ?

➡ **average, mean, mode**

medium
adjective

The medium value is midway between 2 extremes.

(eg) Jason was of medium height.

mega- [**M**]
prefix

Mega- is a prefix meaning 1 000 000.

megabyte [**MB**]
noun

A megabyte is 1,048,576 (2^{20}) bytes, or 1024 kilobytes.

(eg) Computer memory size is measured in kilobytes or megabytes.

➡ **byte, kilobyte**

memory
noun

A computer's memory is its data storage capacity.

eg A computer's random access memory (RAM) is lost when the computer is switched off.

mensuration
noun

The measurement of length, area and volume is called mensuration.

eg Turn to the chapter on mensuration in your textbooks.

i The word mensuration comes from the Latin word *mensura*, meaning size, measure or measurement.

menu
1 noun

A menu in a restaurant is a list of food served there.

1 2 3 4

2 noun

A computer program's menu is the list of options available.

eg A file menu gives options such as 'open a new file', 'save a file' or 'print a file'.

meridian
noun

A meridian is an imaginary line drawn through both the north and the south pole.

eg The longitude 0°, known as the Greenwich meridian, is the meridian passing through Greenwich, England.

➡ **latitude, longitude**

method
noun

? ? ? ?

A method is a description of the way of doing an experiment or process.

eg Simultaneous equations are solved graphically or using the elimination method.

metre [m]
noun

A metre is a measurement of length in the metric system.

eg There are 100 centimetres in 1 metre.

metres per second [m/s]
noun

Metres per second is the number of metres covered in 1 second.

eg The athlete's average speed was 5 metres per second.

metric unit
noun

Metric units are units of weight and measurement in the metric system, a number system based on multiples of 10.

eg Metres and grams are metric units.

i The metric system was developed in France in the 18th century. In 1897 a law was passed in the UK giving permission to use metric weights and measures. SI (System International d'Unites) is the international system of metric weights and measures.

➡ **imperial unit**

micro-
prefix

Micro- is a prefix meaning $\frac{1}{1000000}$.

midday
noun

Midday is 12 o'clock in the middle of the day.

eg In the 12 hour clock, the time from midnight to midday is a.m.

middle
noun

The middle is the centre.

eg The middle of a square is the point where the diagonals intersect.

➡ **median, midpoint**

midnight
noun

Midnight is 12 o'clock in the middle of the night.

eg In the 12 hour clock, the time from midday to midnight is p.m.

midpoint
noun

The point dividing a line into 2 equal parts is called the midpoint.

eg

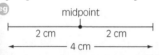

mile
noun

A mile is an imperial measurement of length.

eg There are 1760 yards in 1 mile.

ℹ Mile comes from the Latin word *nille*, meaning one thousand: a mile was defined as the distance of one thousand paces.

miles per hour [mph]
noun

Miles per hour is the number of miles covered in 1 hour.

eg What is the average speed of a car, in miles per hour, that travelled a distance of 35 miles in 30 minutes?

millennium
noun

A millennium is a period of 1000 years.

eg Most people celebrated the end of the last millennium and the beginning of this one at midnight on 31 December 1999. Some people think the celebration should have been on 31 December 2000.

milli-
prefix

Milli- is a prefix meaning $\frac{1}{1000}$.

milligram [mg]
noun

A milligram is a measurement of weight or mass equal to $\frac{1}{1000}$ gram.

eg There are 1000 milligrams in 1 gram.

millilitre [ml]
noun

A millilitre is a measurement of capacity equal to $\frac{1}{1000}$ litre.

eg There are 1000 millilitres in 1 litre.

millimetre [mm]
noun

A millimetre is a measurement of length equal to $\frac{1}{1000}$ metre.

eg There are 10 millimetres in 1 centimetre.

million
noun

A million is the number 1 000 000.

eg A million is 1000 squared.

minimum point
noun

The minimum point on a curve is the bottom of that curve.

eg

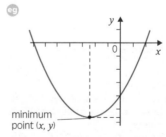

➡ **maximum point**

minimum value
➡ **least value**

minority **1234**
noun

A minority of a group is fewer than half the items or persons in that group.

eg Only a minority of pupils in Year 9 voted to change the colour of their school tracksuit.

➡ **majority**

minor sector
noun

The minor sector is the smaller section of the circle between 2 radii and an arc.

eg

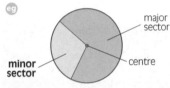

➡ **major sector, sector**

minor segment
noun

The minor segment is the smaller section of the circle between a chord and an arc.

eg

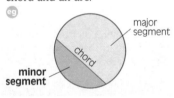

➡ **major segment, segment**

minus [–] **1234**
preposition

You use 'minus' to show that one number is being subtracted from another.

eg What is 12 minus 4?

➡ **subtract, take away**

minute [min] (*min-it*)
noun

A minute is a measurement of time.

eg There are 60 seconds in 1 minute.

minute (*my-newt*)
adjective

Minute means very small.

eg A circle of radius less than 2 mm is minute.

mirror line
noun

A mirror line is another name for a line of symmetry.

eg Jack placed his small mirror on the mirror line. In it he could see the reflection of the parallelogram.

➡ **line of symmetry, reflection**

mistake **? ? ? ?**
noun

A mistake is an error.

eg It is a common mistake to think $2^3 = 2 \times 3$ instead of $2 \times 2 \times 2$.

mixed fraction **1234**
noun

A mixed fraction, or mixed number, is a whole number together with a proper fraction.

eg $9\frac{1}{4}$ is a mixed fraction.

mixed number
➡ **mixed fraction**

Möbius strip (*mer-be-us*)
noun

A Möbius strip is a flat strip which is twisted halfway and the ends joined together. It has only one side and only one edge.

eg

🔢 The Möbius strip was invented in the 19th century by August Möbius, a German mathematician.

modal class
noun

In a grouped frequency distribution, the largest frequency group is the modal class or group.

The modal class was found to be from 1 p.m. to 2 p.m., when most customers came into the store.

modal group
➡ modal class

mode
noun

An average value that is the most frequent value is the mode.

What is the mode of the following: 4, 5, 3, 6, 7, 4, 5, 7, 4?

➡ average, mean, median

month
noun

A month is approximately 4 weeks, the time it takes for the moon to go round the Earth.

There are 12 months in 1 year.

Some months have 30 days, others have 31 days. Only February has 28 days, or 29 in a leap year.

mortgage (*mor-gij*)   **1 2 3 4**
noun

A mortgage is a loan, extended by a building society or bank, for buying a house.

Dennis and Joan applied for a mortgage to the building society.

➡ loan

most significant digit
noun   **1 2 3 4**

When correcting to a given number of decimal places or significant figures, the most significant digit is the leading digit.

25.307 corrected to 3 significant figures is 25.3, with the underlined 2 as

the most significant digit. It holds its place value of 20 and maintains the magnitude of the original number.

➡ least significant digit

multiple $+\sqrt{x}\div$
noun

If one number divides exactly into another number, the second is a multiple of the first.

Every multiple of 5 ends in 0 or 5.

➡ common multiple, lowest common multiple

multiplication [×] $+\sqrt{x}\div$
noun

Multiplication is the operation of adding a number to itself a given number of times.

It helps if you learn your multiplication tables.

multiply [×] $+\sqrt{x}\div$
verb

To multiply is to add a number to itself a given number of times.

Multiply 6 by 4.

multiply out $x = a^2$
verb

Multiplying out brackets, in an algebraic expression or equation, is expanding them.

Multiply out the brackets and solve the equation: $5(y + 2) = 48$.

➡ expand

mutually exclusive
adjective

If A and B are events which cannot happen at the same time, they are mutually exclusive events.

Throwing a 'head' and throwing a 'tail' with the same toss of a coin are mutually exclusive events.

➡ independent, 'or' rule

M
N
O
P

Nn

narrow
adjective

Narrow means thin, of small width.

eg Kim cut a narrow strip, 1.5 cm wide, off the piece of card.

➡ **wide**

natural number
noun

A natural number is a positive integer.

eg The natural numbers are 1, 2, 3, 4, 5…

nautical mile
noun

A nautical mile is a measurement used at sea and equal to 1852 metres.

eg 1 knot is a unit of speed of 1 nautical mile per hour.

near
adjective/adverb

Near is close to or adjacent.

eg Round 159 to the nearest 100.

➡ **adjacent**

negative number
noun

A number less than zero is a negative number.

eg −4.9, −27 and −0.451 are negative numbers.

➡ **directed number, positive number, sign**

net
1 noun

A net is a surface which can be folded into a solid.

eg

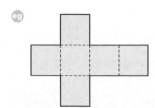

2 adjective

The net amount is that amount left after all deductions.

eg Maya has a gross income of £10,250. Her deductions come to £3,000. What is her net income?

➡ **gross**

network
1 noun

A network is a group of computers connected together.

eg Several users on the same network can access the same data.

2 noun

A network is a set of lines connecting a set of points.

eg The London Underground map is an example of a network.

no chance
➡ **chance**

nonagon
noun

A nonagon is a 9-sided polygon.

eg

north

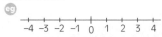

noun

North is one of the 4 main points of a compass, opposite to south.

eg

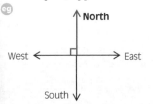

notation $x = a^2$

noun

Notation is the use of different signs and symbols to represent mathematical statements.

eg The notation used for the index or power of a number is a small digit placed at the top right of the number.

➡ algebra, sign, symbol

nth term $x = a^2$

noun

The nth term is the general term of a number sequence.

eg The nth term of the sequence 5, 8, 11, 14, 17... is $3n + 2$.

➡ sequence

number 1234

noun

A number is a symbol used for counting.

eg 1, 2, 3, 4... are natural numbers.

number bond 1234

noun

A number bond is a pair of numbers with a particular total.

eg These number bonds make 10: $1 + 9, 2 + 8, 3 + 7, 4 + 6, 5 + 5$.

number line 1234

noun

A line with a scale, showing numbers in order, is known as a number line.

eg

```
 +--+--+--+--+--+--+--+--+--+
-4 -3 -2 -1  0  1  2  3  4
```

number sentence 1234

noun

A number sentence is a statement used in mathematics which has numbers instead of words.

eg $2 + 3 = 5$ is a number sentence.

number square 1234

noun

A number square is a square grid with cells numbered in order.

eg

1	2	3
4	5	6
7	8	9

numeral 1234

noun

A numeral is a symbol used for a number.

eg The Roman numeral V represents the number 5.

numerator 1234

noun

The numerator is the number above the line in a fraction.

eg In the fraction $\frac{5}{7}$ the numerator is 5.

➡ denominator, fraction

Oo

object
noun

An object is a shape.

eg An object was reflected in its line of symmetry and produced an image.

oblique (ob-**leek**)
adjective

Oblique means sloping or slanted.

eg A diagonal is an oblique line across a shape from vertex to vertex.

oblong
noun

An oblong is a shape with two pairs of straight, unequal sides and four right angles..

eg The lawn was in the shape of an oblong.

➡ **rectangle**

obtuse angle
noun

An obtuse angle lies between 90° and 180°.

eg

138°

➡ **acute angle, reflex angle**

obtuse triangle
noun

An obtuse triangle is a triangle that has an obtuse angle.

eg

135°

➡ **acute triangle**

occur
verb

To occur is to happen.

eg How often did this event occur?

occurrence
noun

An occurrence is an event.

eg A hurricane in Britain is a rare occurrence.

➡ **event**

octagon
noun

An octagon is an 8-sided figure.

eg

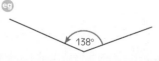

octahedron
noun

An octahedron is an 8-sided solid figure.

eg If a solid shape has 8 faces that are equilateral triangles, it is a regular octahedron.

odd number
noun

When a number is divided by 2 and gives a remainder 1 it is an odd number.

eg 1, 3, 5, 7, 9... are all odd numbers.

➡ **even number**

ogive (oh-*jive*)
noun

An ogive is a cumulative frequency curve in the shape of a shallow 's' or ogee.

eg

➡ **cumulative frequency graph**

open

$x = a^2$

1 adjective

If something is open it is not closed.

eg The sequence of even numbers from 2 to 20 inclusive is closed, but the sequence of even numbers from 2 is open as there is no specified end.

2 verb

➡ expand

operating system

noun

An operating system is the basic software that runs a computer and processes other software.

eg Windows 2000 and Windows XP are both operating systems.

operation

$+\sqrt{x} \div$

noun

A way of combining numbers is known as an operation.

eg The four number operations are addition, subtraction, multiplication and division.

opposite

1 noun

Something that is completely different is the opposite.

eg Addition is the opposite of subtraction.

2 preposition

To be opposite something is to be in a facing position.

eg In an isosceles triangle, the 2 equal angles are opposite the 2 equal sides.

option

noun

An option is a choice, preference or selection.

eg The computer program gave a number of different options.

order

1 noun

An order is a command or instruction.

eg The order given to the computer told it to add the figures in the vertical column of the spreadsheet.

2 noun   1 2 3 4

When you arrange items in ascending or descending sequence, you are putting them in order.

eg Put items in order, starting with the smallest.

➡ ordinal number

order of operations

➡ BIDMAS

order of rotational symmetry

noun

The order of rotational symmetry is the number of positions a shape can take, when rotated, and still look the same.

eg

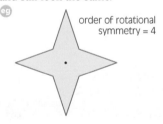

order of rotational symmetry = 4

ordinal number   1 2 3 4

noun

An ordinal number describes a position in a number sequence.

eg First (1st), second (2nd), third (3rd) and fourth (4th) are all ordinal numbers.

➡ cardinal number

organise   1 2 3 4

verb

To organise is to plan something or to arrange it in order.

eg Organise the numbers 1 to 10 into a number square.

origin
$x = a^2$

noun

The origin is the point, with coordinates (0,0), where the *x*-axis and *y*-axis cross.

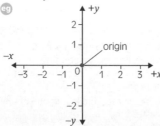

'or' rule

noun

The 'or' rule refers to the probability of a combination of mutually exclusive events, and is expressed as

P(A or B) = P(A) + P(B).

When you are asked to find the probability of event A or event B occurring, use the 'or' rule.

➡ 'and' rule, independent, mutually exclusive

ounce [oz]

noun

An ounce is an imperial unit of mass or weight.

16 ounces = 1 pound

outcome

noun

Outcomes are possible results of an experiment.

P(red Jack from a pack of cards)

$$= \frac{\text{Successful outcomes}}{\text{Total outcomes}} = \frac{2}{52} = \frac{1}{26}$$

outline

1 noun

An outline is the outside edge of a drawn shape or sketch.

The outline of a polygon is its perimeter.

2 verb
? ? ? ?

To outline a plan or idea is to explain it briefly.

Outline the task before you decide on the method.

output

noun

The amount produced by a machine or process is the output.

Computer output can be viewed on screen or printed out.

outstanding
1 2 3 4

adjective

Bills left unpaid are said to be outstanding.

After Fiona had paid the deposit for the microwave oven, the outstanding balance was £100.

oval

noun

An egg-shaped object is an oval.

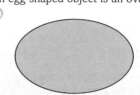

➡ ellipse

owe
1 2 3 4

verb

If something is owed it has not been returned or repaid.

There is only one instalment owing.

➡ debt, lend, loan

P(A)
noun

P(A) is the symbol for the probability of an event A occurring.

🔲 P(red card from a pack of cards) $= \frac{26}{52} = \frac{1}{2}$

➡ **probability**

pack
noun

A pack of playing cards is a pack of 52 cards used for playing a variety of games.

🔲 If you pick a card from a pack, what is the probability that is an ace?

ℹ A pack of playing cards has 4 suits – spades, hearts, diamonds and clubs. Spades and clubs are black, diamonds and hearts are red. Each suit has 13 cards: an ace (value 1), cards 2 to 10, followed by 3 picture cards – jack, queen and king.

➡ **suit**

pair
noun

A pair is a set of 2 things.

🔲 A trapezium has 1 pair of parallel lines.

➡ **couple**

palindrome
noun

A palindrome is a number that is the same when reversed.

🔲 The numbers 2002, 13531 and 373 are palindromes.

ℹ The word 'palindrome' comes from the ancient Greek word *palindromos*, meaning 'running back again'. The term also refers to reversible words such as 'level'.

parabola
$x = a^2$
noun

A parabola is a curved graph drawn from a quadratic equation.

🔲

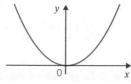

➡ **quadratic equation**

parallel
adjective

Lines that are parallel always stay the same distance apart and never meet.

🔲 Corresponding angles and alternate angles are formed by a line crossing 2 or more parallel lines.

parallelogram
noun

A parallelogram is a quadrilateral with 2 pairs of parallel sides. Its sides are consequently equal in length and the diagonals bisect each other.

🔲

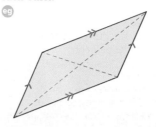

part
noun

A part is a separate piece or division.

🔲 The sweets were divided in the ratio 3:4. There were a total of 7 parts.

A B C D E F G H I J K L M N O

P

Q R S T U V W X Y Z

partition
1234
noun

A partition is a division separating parts of a whole.

🔸 The decimal point acts as a partition between the whole and fractional parts of a number.

Pascal's triangle
1234
noun

Pascal's triangle is a triangular array of numbers with apex 1 and each row starting and ending with 1.

🔸
```
          1
        1   1
      1   2   1
    1   3   3   1
  1   4   6   4   1
1   5  10  10   5   1
```

🔹 Pascal's triangle was devised by the French mathematician Blaise Pascal (1623–1662). It is used to calculate probabilities. Each number is found by adding the 2 numbers above it, and the sum of each row is a power of 2.

past

1 noun

The period of time that is before the present is called the past.

🔸 In the past, when there were no computers, calculations took longer.

2 adjective

Past events have happened before the present.

🔸 The past 3 months had above average rainfall.

3 preposition/adverb

Anything moving ahead of something else is moving past.

🔸 The leading runner went past the finishing line 40 secs ahead of the other runners.

pattern
? ? ? ?
noun

A pattern is an arrangement of numbers or shapes satisfying a given rule.

🔸 A tessellation is a regular pattern of shapes.

pay
1234
noun

Pay, or a payment, is reward for work done.

🔸 Marie works 30 hours at £5.50 an hour. What is her total pay?
➡ **earn, salary, wage**

payment
➡ **pay**

penny (plural: pence) [**p**]
noun
1234

Penny is a unit of money in the UK, as well as many other currencies.

🔸 100 pence = £1

🔹 Before decimal currency was introduced in the UK in 1971, 12 pence (12d) = 1 shilling (1s), and 20 shillings = £1. For some years after 1971, decimal pence were known as new pence.
➡ **pound**

pension (*pen-shn*)
1234
noun

A pension is a regular allowance paid to a retired person.

🔸 The weekly pension for a retired man is approximately £75.

pentagon
noun

A pentagon is a 5-sided figure.

🔸

pentahedron
noun

A pentahedron is a 5-sided solid figure.

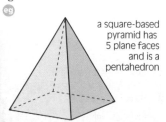

a square-based pyramid has 5 plane faces and is a pentahedron

per
preposition

$+\sqrt{x}-\div$

Per is a Latin word used to mean 'each'.

Kilometres per hour (kph) means the number of kilometres covered in each hour.

per cent [%]
noun

Per cent is another way of writing a fraction with denominator 100.

The interest rate increased to 10 per cent.

The abbreviation % may have been developed in counting houses in order to save parchment by just writing 1 line instead of 2, e.g. $\frac{75}{100}$ could be written as 75 with 100 alongside: 75%.

percentage
noun

Percentage means the proportion or rate per 100 parts.

Kings, queens and jacks form what percentage of a pack of cards?

perfect number
noun

A number that is the sum of its factors is said to be a perfect number.

6 is a perfect number, as $1 + 2 + 3 = 6$.

perimeter
noun

The perimeter of an enclosed area is the boundary or edge of that area, as well as the length of that boundary.

What is the perimeter of a rectangle with length 15 cm and breadth 7 cm?

period
noun

A span or interval of time is called a period of time.

A decade is a period of 10 years.

perpendicular
adjective

A perpendicular line or plane is one at right angles to another line or plane.

perpendicular bisector
noun

A perpendicular bisector is drawn at right angles to the midpoint of a line.

→ bisect, bisector

pi [π]
noun

In a circle, pi is the ratio of its circumference to its diameter.

Pi is 3.14, 3.142 or $\frac{22}{7}$.

pictogram
noun

A chart using pictures to represent numbers of items is a pictogram, or pictograph.

eg

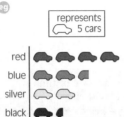

represents	5 cars

red
blue
silver
black

number of cars

pictograph
➡ pictogram

pie chart
noun

Statistical data is sometimes illustrated in a circular chart called a pie chart.

eg

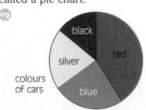

colours of cars

black
red
silver
blue

pint [pt]
noun

A pint is an imperial unit of capacity.

eg 8 pints = 1 gallon

place value
noun

1234

The place value of a digit is its value in relation to its place within the given number.

eg In the number 167.234 the place value of 1 is 100, and the place value of 2 is $\frac{2}{10}$.

plan
noun

A plan is a scale drawing or design for an object or building.

eg A plan of the window design was drawn to a scale of 5 cm to 1 m.

plane
noun

A plane is a flat surface.

eg A line joining any 2 points lies on a flat or plane surface.

plane figure
noun

A plane figure is a 2D shape having length and width but no depth.

eg A rectangle is an example of a plane figure.

plane of symmetry
noun

A plane of symmetry cuts a solid object into half.

eg

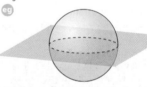

plane section
➡ section

plane symmetry
noun

If a solid object has reflective symmetry, it is said to have plane symmetry.

eg

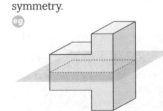

plan view
noun

A plan view is a view of a shape or object when looking downwards onto it.

eg ground floor of a house

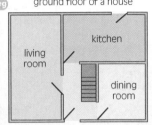

living room

kitchen

dining room

plot
verb $x = a^2$

To plot points for a graph, according to given coordinates, means to mark their position on graph paper.

eg What shape do you find when you plot the points (2,3), (2,6) and (7,3)?

plus [+]
preposition $1\,2\,3\,4$

Plus means in addition to.

eg 14 plus 15 makes 29.

➡ minus

p.m.
noun

The abbreviation for 'post meridiem' is p.m., which means after midday or noon.

eg The train left the station at 2.45 p.m.

➡ a.m.

point
noun $1\,2\,3\,4$

A point is a dot marking position.

eg The line was drawn through each point.

➡ decimal point

pole
noun

The north and south poles are at either end of the Earth's axis.

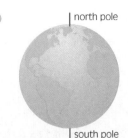

eg north pole

south pole

poll
noun

A poll is a survey of people's intentions or opinions.

eg The poll showed that the government had a lead of 13% over the opposition.

polygon
noun

A polygon is a flat or plane shape with many sides.

eg If a polygon has equal sides it is said to be a regular polygon.

polyhedron (*pol-lee-**hee**-dron*)
noun

A polyhedron is a many-sided solid figure.

eg

poor chance
➡ chance

population
1 noun

The number of people who live in a certain place is its population.

eg The population of the UK is approximately 60 000 000.

population pyramid – prepare

2 noun
A population can refer to any large group of items being investigated.

eg A sample was taken from the larger population of objects.

population pyramid
noun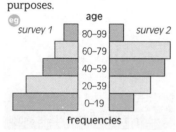
A population pyramid represents 2 sets of statistical data about population for comparison purposes.

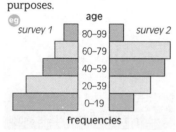

portion 1234
noun
A portion is a part or piece of a whole.

eg If an apple pie is divided into 5 portions, each portion is $\frac{1}{5}$ of the whole.

position
noun
A position is a place.

eg The sum is arranged so that the numbers are in the correct position in relation to the decimal point.

positive number 1234
noun
A positive number is a number greater than zero.

eg $x > 0$ means x is a positive number.
➡ directed number, negative number

possible
adjective
If an event is possible it means it can happen.

eg It is possible that the sun will shine for at least 3 days next week.

pound [lb] [£]
1 noun
A pound (lb) is an imperial unit of weight or mass.

eg There are 14 pounds in 1 stone.
2 noun 1234
A pound (£) is also a unit of money.

eg There are 100 pence in 1 pound.

power
➡ index

practice ????
noun
Practice is an exercise which practises a skill.

eg You need some practice in drawing curved graphs.

practise ????
verb
To practise a skill is to do it regularly.

eg They practised their multiplication tables every day.

precise 1234
adjective
To be precise is to be exact and correct.

eg Be precise and give an accurate answer.
➡ exact

predict
verb
If you predict, you forecast that an event will occur.

eg Emily predicts that she will toss 50 heads and 50 tails in her experiment with flipping a coin.

prepare 1234
verb
To prepare to do something is to get ready for it.

eg Prepare your scale drawing by making a rough sketch.

present

1 noun

The present is today, now or at this time.

eg If Wednesday is the present, last Tuesday was the past.

2 verb **? ? ? ?**

To present something is to perform it or give it to someone else.

eg The maths prize, presented to the top student, was a geometry set.

pressure
noun

Pressure is the force per unit area when something presses on something else.

eg Pressure is measured in pounds per square inch (lb/in²).

previously **1234**
adverb

Something that happened previously occurred earlier.

eg The triangle, which had previously been reflected in the *x*-axis, was now rotated about the origin.

price **1234**
noun

The price of something is its cost or expense.

eg The TV set, which cost £559, was discounted 20% in the sale. What was its sale price?

primary source
noun

Primary source refers to data that is collected to use for investigation.

eg The class collected colours and makes of cars, which was their primary source of data.

➡ **secondary source**

prime factor **1234**
noun

A prime factor is a factor that is a prime number.

eg Express 28 as a product of its prime factors.

➡ **factor**

prime number **1234**
noun

A prime number has only 2 factors, itself and 1.

eg 2, 3, 5, 7, 11, 13, 17 and 19 are all prime numbers.

principal

1 noun **1234**

The principal is the original amount of money invested or borrowed, before interest is calculated.

eg Hayley invests a principal of £250 in the building society at a rate of 5% per annum.

➡ **interest**

2 adjective

Principal means main, largest or most important.

eg The modal group is the principal group in the frequency distribution.

prism (*priz-um*)
noun

A prism is a 3D shape with a uniform cross-section.

eg

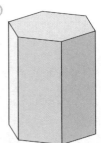

hexagonal prism

probability
noun

The probability of an event occurring is the chance it may happen, which can be given as a fraction, a decimal or a percentage. Probability = number of successful events ÷ total number of possible events

eg What is the probability of throwing a dice and getting a 6?

➔ **experimental probability, P(A)**

probability scale
noun

A probability scale is an ordered line numbered from 0 (the probability of an impossible event) to 1 (the probability of a certain event).

eg

less likely ← → more likely

0   0.5   1
impossible   certain

probable
adjective

If an event is probable it is reasonable and likely.

eg It is probable that the temperature will be less than 12°C on Christmas Day.

problem
noun

A problem is a puzzle or a question.

eg Work through the problem carefully.

procedure
noun

A procedure is a method, a way of doing something.

eg What is the procedure for reducing fractions?

produce
verb

To produce is to create or make something.

eg The factory produced 100 batteries per hour.

product
1 noun

A product is something that is produced or manufactured.

eg The product was exported to France.

2 noun

When 2 or more numbers are multiplied together, the result is called the product.

eg The product of $2 \times 3 \times 4$ is 24.

profit
noun

In business, to make a profit is to make money on a deal.

eg How much should he sell the pens for in order to make a profit?

➔ **loss**

program
noun

A program is a series of instructions to a computer to carry out a procedure.

eg MS Word and Excel are examples of programs.

programme
1 noun

A programme is a plan of action.

eg The programme for the school trip allowed for 2 hours in the museum.

2 noun

A programme is a radio or TV broadcast.

eg The TV programme lasted 45 minutes.

project
1 noun

A project is a task which has to be planned before it is undertaken.

eg Stephen decided to do a Handling Data project.

2 verb

To project a shape is to extend it.

eg The roof was extended to project a distance of 60 cm beyond the wall.

projection

noun

A projection is a mapping of points from a 3D figure onto a line or plane.

eg An atlas is a book of projections of different countries, called maps.

proof

$x = a^2$

noun

A proof is an argument or explanation that establishes the truth of a proposition.

eg The proof concluded that the hypothesis was correct.

proper fraction

1 2 3 4

noun

A proper fraction has a numerator smaller than its denominator.

eg $\frac{12}{13}$ is a proper fraction.

➡ improper fraction

property

? ? ? ?

noun

Property is something that belongs to someone.

eg You can protect your property by taking out insurance.

proportional

1 2 3 4

adjective

If quantities vary according to a given ratio, they are said to be proportional, or have proportionality.

eg Corresponding sides of similar figures are proportional.

protractor

noun

A protractor is an instrument used for measuring angles.

eg

prove

$x = a^2$

verb

To prove is to establish the truth of a proposition through a chain of reasoning.

eg Prove that the triangle has a right angle by using Pythagoras' theorem.

purchase

➡ buy

pyramid (*pi-ram-id*)

noun

A pyramid is a solid shape with triangular faces meeting at a vertex.

eg Pyramids are named after the shape of the base. A square-based pyramid, for example, has a base in the form of a square.

Pythagoras' theorem

(*pie-**thag**-or-us*)

noun

Pythagoras' theorem states that in a right-angled triangle the square on the hypotenuse is equal to the sum of the squares on the other two sides.

eg

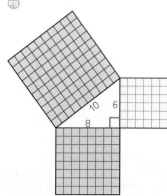

Pythagoras was a Greek mathematician who lived in the 6th century BC.

➡ hypotenuse

Qq

quad-
1 2 3 4
prefix
Quad- is a prefix meaning 4.

quadrangle
noun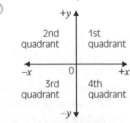
A 4-sided area enclosed by walls or buildings is known as a quadrangle.

eg What is the perimeter of a rectangular quadrangle with 2 sides of 20 metres and 2 sides of 16 metres?

quadrant
1 noun
A quadrant is quarter of a circle.
eg

2 noun
A quadrant is also one of the 4 regions of a plane that is divided by the x-axis and y-axis.
eg

	$+y$	
2nd quadrant		1st quadrant
$-x$	0	$+x$
3rd quadrant		4th quadrant
	$-y$	

quadratic equation
noun $x = a^2$
A quadratic equation is an equation containing unknowns to maximum power 2. It has 2 solutions.

eg The quadratic equation $3x^2 - 7x + 2 = 0$ has solutions $x = \frac{1}{3}$ or $x = 2$.

quadratic expression
noun $x = a^2$
A quadratic expression is an expression containing unknowns to maximum power 2.

eg $ax^2 + bx + c$ is a general quadratic expression.

quadratic function $x = a^2$
noun
A quadratic function is a relationship between 2 variables.

eg y as a quadratic function of x is shown in $y = x^2 + 2$.

quadratic sequence
noun **1 2 3 4**
In a quadratic sequence the difference between the terms changes by the same amount each time. Each term is compared with the square numbers 1, 4, 9, 16, 25…

eg To find later terms of the quadratic sequence 3, 6, 11, 18, 27…:

1^2	2^2	3^2	4^2	5^2
3	6	11	18	27
$2+1^2$	$2+2^2$	$2+3^2$	$2+4^2$	$2+5^2$

so nth term is $2+n^2$.

quadrilateral
noun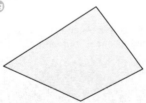
A quadrilateral is a 4-sided polygon.
eg

➡ **kite, parallelogram, rectangle, rhombus, square, trapezium**

quadruple
1 2 3 4
adjective

4 times a number is its quadruple.

eg Wasim scored quadruple points for getting that answer.

quantity

noun

A quantity is an amount.

eg If a pudding for 4 people needs 60 g of sugar, what quantity of sugar is needed for 2 people?

➡ **amount**

quarter
1 2 3 4
noun

A quarter is one of 4 parts.

eg

$\frac{1}{4}$	$\frac{1}{4}$
$\frac{1}{4}$	$\frac{1}{4}$

quartile
noun

A quartile is one of 3 values which divides a frequency distribution into 4 intervals. The lower quartile is at $\frac{1}{4}$ level, the median is halfway and the upper quartile is at $\frac{3}{4}$ level.

eg

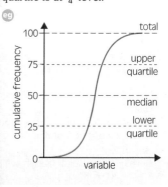

➡ **cumulative frequency graph, interquartile range**

question
noun

A question is something that is asked.

eg There were 20 questions in the test.

questionnaire
noun

A questionnaire is a sheet of questions used to collect data.

eg All questions on a questionnaire must be fair and not biased.

quotient (***kwo***-shnt)   +√×−÷
noun

A quotient is the result of dividing one number by another.

eg What is the quotient when you divide 248 by 4?

Rr

radial
adjective

A radial arrangement has a centre with ray-like components, as in a bicycle wheel.

eg

radian
noun

A radian is a unit of measurement of an angle, equivalent to 57°.

eg There are 2π radians in 1 revolution (360°).

radius (plural: radii)
noun

A radius is a line joining the centre of a circle or sphere to a point on the circumference.

eg

radius

raise
verb

To raise something is to take it to a higher level, including raising a term involving a power to another power.

eg $(a^2)^3$ means a^2 is raised to the power 3: $(a^2)^3 = a^{(2\times3)} = a^6$.

random
adjective

Something that happens by chance or without bias is random.

eg A random selection of coloured balls was drawn from the bag.

random sample
noun

A random sample is a selection where each sample has an equal chance of being selected.

eg Jessica needed to question a random sample of shoppers for her survey.

➡ **representative sample**

range
noun

The range is the spread of data; it is the difference between the greatest and least values.

eg The top mark was 75% and the bottom was 22%, so what was the range of marks?

rank
verb

If items are ranked, they are set in certain positions.

eg The students' marks were ranked in ascending order of their grades.

rate
1 *noun*

The rate of something is the ratio of one quantity to a unit of another.

eg Speed is the rate of distance covered in a given unit of time.

2 *noun*

Interest is charged at a given rate.

eg The building society changed its interest rate from 3.35% to 2.85%.

ratio (*ray-she-oh*) 1 2 3 4
noun

A ratio is used to give a part-to-part comparison.

eg The school shop sold 150 cans of drink in the ratio of 3:2:5 for lemonade, orangeade and cola. How many cans of each were sold?

raw data
noun

Unprocessed data is called raw data.

eg The number of people watching different sports was still raw data as it had not yet been organised into groups.

rear
adjective

Rear refers to the back of an object.

eg Draw the rear elevation of the house.

reason ? ? ? ?
noun

A reason is an explanation for something.

eg The reason that the angles were equal was that they were vertically opposite angles.

reasonable
adjective

A reasonable statement or result is one that makes sense.

eg Is it reasonable to say that you need 30 cans of paint to decorate the kitchen?

reciprocal 1 2 3 4
noun

A reciprocal is the inverse of any number except zero.

eg $\frac{1}{4}$ is the reciprocal of 4.
➡ **inverse**

record
verb

To record something is to write it down.

eg The results of the experiment were recorded on a spreadsheet.

rectangle
noun

A rectangle is a quadrilateral with 2 pairs of opposite, equal parallel sides and the diagonals bisecting each other; each angle is 90°.

eg What is the area of a rectangle with length 8.4 cm and width 4.8 cm?

recurring 1 2 3 4
adjective

A recurring decimal has digits that are in a continuous pattern, like 0.3333… or 0.252525…

eg 0.3 recurring can be written as 0.$\dot{3}$.
➡ **terminating**

reduce 1 2 3 4
verb

1 To reduce is to make smaller in size.

eg Reduce the size of the rectangle by a scale factor of $\frac{2}{3}$.

2 A fraction is reduced by cancelling to its lowest terms.

eg Reduce $\frac{24}{36}$ to its lowest terms.
➡ **lowest terms, simplest form**

reflection
noun

A reflection is the image seen in a mirror, or produced by reflecting an object in an axis of symmetry.

eg

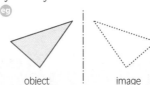

object   image

reflection symmetry
➡ **line symmetry**

reflex angle
noun

A reflex angle is an angle that lies between 180° and 360°.

eg

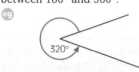

320°

➡ **acute angle, obtuse angle**

region $x = a^2$
noun

A region is an area or part of a quadrant bounded by given lines.

eg

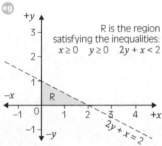

R is the region satisfying the inequalities:
$x \geq 0$ $y \geq 0$ $2y + x < 2$

$2y + x = 2$

regular
1 adjective

A regular polygon has sides of equal length and is equiangular.

eg A regular pentagon has exterior angles of 72°.

2 adjective

Events that happen at regular intervals occur at equal time periods.

eg We hold regular meetings on the first Wednesday of the month.
➡ **irregular**

relate   ? ? ? ?
verb

To relate is to make a connection between 2 or more items.

eg The sides of a right-angled triangle are related to one another in the trigonometrical ratios cosine, sine and tangent.

relative frequency
noun

Relative frequency is a way of estimating probabilities if they cannot be accurately calculated.

eg Relative frequency = (the number of times an event has occurred) ÷ (the total number of experiments)

remainder $+\sqrt{x}-\div$
noun

When a quotient has to be a whole number, the amount left over is called the remainder.

eg $29 \div 3 = 9$ remainder 2
➡ **division, quotient**

rent $1\,2\,3\,4$
noun

Rent is the money you pay to the owner, or landlord, of a property in which you live.

eg One of Jane's biggest expenses was the rent of her flat, which was £360 per month.

repeated addition $x = a^2$
noun

Repeated addition or subtraction is the process of continuing to add or subtract the same amount.

eg The sequence 3, 7, 11, 15… is generated by repeated addition of 4.

repeated subtraction
➡ **repeated addition**

replace $x = a^2$
verb

To replace is to put something in place of, or to substitute.

eg What is the value of $\frac{1}{2}bh$ when you replace b with 7 and h with 4.8?

represent
verb

Something that symbolises something else represents it.

eg In the formula πr^2, 'r' represents the radius.

representative sample
noun

A representative statistical sample accurately shows the general population being investigated.

eg A representative sample is accurate and unbiased.

➡ random sample

resources
noun

Resources are the materials needed for a project.

eg The resources needed by the class for building models of solid figures included card, scissors and glue.

result
noun

The result is the answer.

eg What is the result if you increase the speed by 50%?

retail price
➡ selling price

reverse
noun

The reverse of something is its opposite, when the thing is turned back to front.

eg The reflected image is the reverse of the original object.

revolution
noun

A revolution is a complete turn through 360°.

eg 4 right angles make 1 complete revolution.

rhombus (*rom-buss*)
noun

A rhombus is a parallelogram with 4 equal sides but no right angles. Other terms for rhombus are lozenge and diamond.

eg

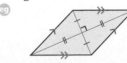

➡ parallelogram

right
adjective

To be right is to be correct.

eg Substitute the solutions in one of the pair of simultaneous equations to check if they are right.

right angle [⌐]
noun

A right angle is a quarter of a revolution, or 90°.

eg A square has 4 right angles.

right-angled triangle
noun

A right-angled triangle has 1 angle of 90°.

eg

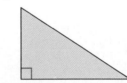

risk
noun

Risk means chance or probability.

eg The risk of it raining on any day in June in Brighton is $\frac{2}{7}$.

Roman numerals
noun

Roman numerals are a number system used by the ancient Romans.

eg 1902 is MCMII in Roman numerals.

➡ C, D, I, L, M, V, X

root [√]

1 2 3 4

noun

A root is a quantity that when multiplied by itself a certain number of times (twice for a square root, three times for a cube root) equals a given quantity.

🔵 $\sqrt[n]{x}$ is the *n*th root of *x*.

➡ **cube root, square root**

rotate

verb

To rotate is to turn round.

🔵 When a shape is rotated, it is turned around a centre of rotation and through a given angle, either clockwise or anticlockwise.

➡ **centre of rotation**

rotation symmetry

noun

A shape has rotation symmetry if there are a number of positions the shape can take, when rotated, and still look the same.

🔵

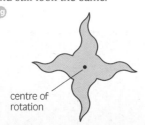

centre of
rotation

rough

1 2 3 4

adjective

A rough calculation is not exact but approximate.

🔵 Give a rough estimate for 299 × 41.

round

1 adjective

Round means circular.

🔵 CD-ROMs have a round shape.

2 verb

1 2 3 4

To round a number is to express it to a required degree of accuracy.

🔵 Round 17.537 (1) to the nearest whole number; (2) to 3 significant figures; (3) to 2 decimal places.

row

noun

A row is a horizontal arrangement.

🔵 A spreadsheet has rows and columns forming cells into which data are entered.

➡ **column**

rule

1 2 3 4

noun

A rule is a given method for a procedure.

🔵 The rule for reducing fractions to their lowest terms is to divide the numerator and denominator by the highest common factor.

ruler

noun

A ruler is an instrument used to draw and measure a straight line.

🔵

salary

1 2 3 4

noun

A salary is a fixed regular payment earned by an employee each month or year.

eg Her salary of £15,600 per year gave her a monthly income, before deductions, of £1,300.

! The word salary comes from a Latin word, *salarium*, meaning money used to buy salt.

➡ **earn, wage**

sale price

+√x−÷

noun

The sale price is the price of an item when it is sold at a discount.

eg What is the sale price, at a discount of 20%, of a ladder costing £75?

➡ **cost price, discount**

sample

noun

A sample is a section of a population or a group of observations.

eg If we ask every tenth person, this will give us a representative sample.

➡ **random sample, representative sample, survey**

sample space

noun

A sample space contains all possible outcomes of an experiment.

eg The sample space for an experiment tossing 2 coins is:

	H	T
H	HH	HT
T	TH	TT

satisfy

$x = a^2$

verb

If a rule is satisfied by a variable, the variable obeys the rule.

eg Do the coordinates (3,4) satisfy the equation $y = 5x - 11$?

scale

1 noun

A scale is a marked measuring line.

eg

scale

2 noun

A scale is also a graded system of measurements.

eg We use the Celsius scale in preference to the Fahrenheit scale.

3 noun

A scale for a drawing or map is the ratio of the drawn distance to its true value.

eg According to the scale, 1 cm represents 5 km.

➡ **scale drawing**

scale drawing

noun

A scale drawing is a diagram drawn to a given scale.

eg Find the distance of the boat from the port using a scale drawing.

➡ **scale**

scale factor

noun

A scale factor for an enlarged figure is the ratio of the enlarged distance to the corresponding original value.

eg If a figure is enlarged by a scale factor of 2, it doubles in size.

➡ **enlargement**

scalene triangle
noun

A scalene triangle has no equal sides or angles.

eg

scales
noun

Scales are an instrument for weighing.

eg

→ **balance**

scatter diagram
→ **scatter graph**

scatter graph
noun

A scatter graph, or scatter diagram, compares 2 variables by plotting 1 value against the other.

eg

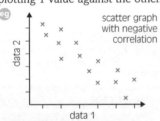

data 2

scatter graph with negative correlation

data 1

→ **correlation, line of best fit**

schedule (*shed-yool*)   ? ? ? ?
noun

A schedule is a timetable or a timed plan of procedure for a project.

eg Mark on your schedule when you propose to begin writing up the project.

score   1 2 3 4
1 noun

A score is 20.

eg He lived 3 score years and 10.

The word 'score' comes from old Saxon 'sceran' meaning 'to cut'. It is thought that, in medieval times, a notch was cut into a tally stick each time a group of 20 was recorded.

2 verb

You score in a game, competition or sport if you earn points or get goals.

eg Billy won the tennis match with a score of 6–4, 6–2, 6–3.

second [2nd] [sec]
1 adjective   1 2 3 4

The ordinal number after first is second (2nd).

eg February is the second month of the year.

2 noun

A second is a unit of time (sec).

eg 60 seconds = 1 minute

→ **hour, minute, time**

secondary source
noun

Secondary source refers to data used for investigation after it has been collected by another person

eg Colin used a secondary source of data taken from government statistics.

→ **primary source**

second difference
→ **first/second difference**

section
noun

A section, or plane section, is formed when you cut a solid figure with a plane.

eg The plane cut through the cube at an angle, forming a triangular section.

→ **cross-section**

sector
noun

A sector is a section of a circle between 2 radii and an arc.

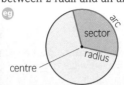

➡ **major sector, minor sector**

segment
1 noun

A segment is a part of a line.

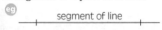

segment of line

2 noun

A segment is also a section of a circle between a chord and an arc.

➡ **major segment, minor segment**

self-inverse $x = a^2$
adjective

A self-inverse function reverses the original mapping. The function $x \rightarrow c - x$ is self-inverse.

When $x = 1$ is substituted in the self-inverse function $x \rightarrow 7 - x$, the result is 6. This is reversed by substituting $x = 6$ in $x \rightarrow 7 - x$. The result is 1, the original value of x.

selling price 1234
noun

The price of an item when it is sold is the selling price, or retail price.

The selling price of the dishwasher is £399.

semicircle
noun

A semicircle is half a circle.

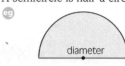

diameter

➡ **circle**

separate $x = a^2$
verb

To separate is to set apart, or divide into component parts.

Separate the terms by using brackets.

sequence $x = a^2$
noun

A collection of terms following a rule or pattern is called a sequence, or series.

What are the next 2 terms of this sequence: 5, 6, 8, 11...?

series
➡ **sequence**

service charge 1234
noun

A service charge, or standing charge, is the fixed amount put on a bill before units are calculated.

The service charge on your gas bill remains at £12.

set square
noun

A set square is an instrument, in the form of a right-angled triangle, which is used for drawing parallel lines.

shape
noun

A shape is a figure made up of drawn lines.

eg A circle is a round shape.

share
1 noun

1234

A share of something is a portion.

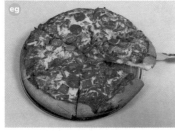

2 verb

To share is to divide equally between 2 or more people.

eg If 24 sweets are shared between 3 people, each person has 8 sweets.

➡ **divide**

shortage
noun

1234

A shortage is a lack or scarcity of something.

eg There was a shortage of protractors, so only some of the class could measure their angles.

side
noun

A side of a figure is the line that forms part of the perimeter or edge.

eg A square has 4 equal sides.

sigma [Σ]
noun

Sigma is the Greek letter used to signify the sum of a set or sequence of numbers.

eg Estimate of mean = $\Sigma fx \div \Sigma f$.

➡ **estimate of mean**

sign
1 noun

$+\sqrt{\times}-\div$

A sign is a symbol that shows the operation to be used.

eg The addition sign is + and the multiplication sign is ×.

2 noun

Sign also refers to the positive (+) or negative (−) value of a directed number.

eg $(-5)^2 = 25$, which has a positive sign because $- - = +$.

➡ **sign rule**

sign change key
noun

1234

The sign change key on a calculator changes positive values to negative values or vice versa.

eg The sign change key has the symbol +/−.

significant
adjective

Something that is significant is important.

eg The results from the experiment were significant as they pointed to a definite connection between the test marks for English and French.

significant figures
[**sf, sig.fig.**]
noun

1234

Significant figures are the number of digits in a number giving a required degree of accuracy.

eg $0.005206 = 0.00521$ to 3 significant figures.

sign rule
noun

1234

Sign rules are rules used with directed numbers.

eg The sign rules are:
(1) $+ + = +$;   (2) $- - = +$;
(3) $+ - = -$;   (4) $- + = -$.

➡ **directed number**

similar figure
noun

Figures having the same shape, but not the same size, are said to be similar figures.

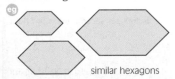

similar hexagons

simple fraction
➡ **common fraction**

simple interest $1\,2\,3\,4$
noun

Simple interest is calculated on the principal alone.

🇪🇬 If £500 is invested at 5% per annum, how much will this total after 3 years' simple interest?

➡ **compound interest, interest**

simplest form $x=a^2$
noun

A fraction is in its simplest form when it cannot be reduced further.

🇪🇬 $\frac{1}{3}$ is in its simplest form.

➡ **lowest terms, reduce**

simplify $x=a^2$
verb

To simplify something is to make it easier to understand.

🇪🇬 Simplify the algebraic expression by collecting like terms.

➡ **cancel**

simultaneous equation
noun $x=a^2$

Simultaneous equations are 2 linear equations satisfied by the same pair of values.

🇪🇬 Simultaneous equations can be solved by the elimination method or graphically. The solutions are the coordinates of the intersection of the 2 lines.

➡ **equation, linear equation**

sine [sin] $1\,2\,3\,4$
noun

The sine of an acute angle, in a right-angled triangle, is the ratio of the side opposite the given angle to the hypotenuse:

🇪🇬
$$\sin X = \frac{\text{opposite}}{\text{hypotenuse}}$$

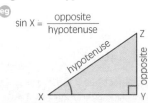

➡ **cosine, tangent, trigonometry**

single $1\,2\,3\,4$
adjective

A single item is one that is on its own, separate from all the others.

🇪🇬 A single point separated the two teams.

sketch
noun

A sketch is a rough drawing.

🇪🇬 Draw a sketch before doing a scale drawing.

slant height
noun

The slant height is the height of a shape, such as a cone, measured along the sloping side.

🇪🇬

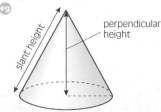

perpendicular height

slope
noun

A slope is an incline.

🇪🇬 The measure of the steepness of a slope is the gradient.

➡ **gradient, incline**

SOHCAHTOA

noun

SOHCAHTOA is an acronym which helps you remember trigonometrical ratios.

eg SOHCAHTOA stands for: **S**ine **O**pposite **H**ypotenuse **C**osine **A**djacent **H**ypotenuse **T**angent **O**pposite **A**djacent.

solid figure

noun

If an object has 3 dimensions it is known as a solid figure or solid shape.

eg A cuboid is a solid figure.

solid shape

➡ solid figure

solution

$x = a^2$

noun

The solution to a problem is the result.

eg What is the solution to the equation $2(y + 4) = 12$?

south

noun

South is one of the 4 main points of a compass, opposite to north.

eg

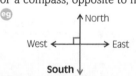

space

noun

The space in a solid figure is the area that it contains.

eg Capacity is the amount of space in a container.

➡ capacity

speed

noun

Speed is the rate at which an object moves relative to time.

eg The train travelled at a speed of 70 mph.

➡ metres per second, miles per hour

sphere

noun

A sphere is a circular solid shape in which all the points on its surface are equidistant from the centre.

eg

spinner

noun

A spinner is a top with numbers marked on it.

eg What is the probability that a spinner marked with the numbers 1 to 6 lands on a 5?

spread

noun

The spread of a frequency distribution is the way it is distributed about the mean.

eg The values were evenly spread on either side of the mean.

➡ range

spreadsheet

noun

A computer program that produces sheets of paper divided into rows and columns is called a spreadsheet.

eg One spreadsheet program is Microsoft Excel.

square

1 noun

A square is a quadrilateral having 4 equal sides and 4 right angles.

eg

2 verb **1 2 3 4**

To square a number is to multiply it by itself.

eg 10 squared (10^2) is 10×10.

➡ square number

square-based pyramid
➡ pyramid

square centimetre [cm^2]
noun

A square centimetre is a unit of area, a square measuring 1 cm × 1 cm.

eg How many square centimetres are there in a square metre?

square kilometre [km^2]
noun

A square kilometre is a unit of area, a square measuring 1 km × 1 km.

eg How many square metres are there in a square kilometre?

square metre [m^2]
noun

A square metre is a unit of area, a square measuring 1 m × 1 m.

eg How many square metres are there in a shape measuring 10 m × 6 m?

square millimetre [mm^2]
noun

A square millimetre is a unit of area, a square measuring 1 mm × 1 mm.

eg How many square millimetres are there in a square centimetre?

square number [n^2] **1 2 3 4**
noun

A square number, or a number squared, is the product of 2 equal factors.

eg Square numbers are always positive: $+a \times +a = a^2$ and $-a \times -a = a^2$.

➡ square

square root [$\sqrt{\ }$] **1 2 3 4**
noun

The square root of a certain number is the number that, when squared, gives that certain number.

eg The square root of 49 is 7.

➡ cube root, root

standard form **1 2 3 4**
noun

Standard form, or standard index form, is a shorthand way of writing very small or very large numbers; these are given in the form $a \times 10^n$ where a is a number between 1 and 10.

eg 2 150 713 in standard form = $2.150\,713 \times 10^6$.

➡ index, power

standard index form
➡ standard form

standard unit **1 2 3 4**
noun

A standard unit is a unit of measurement that a community agrees to use.

eg Examples of standard units in the metric system are the metre and the kilogram.

standing charge
➡ service charge

statement

noun

Information given in the form of words or mathematical symbols is a statement.

eg The following is a statement: 'The cube has 6 faces and 12 edges, all of 5 cm.'

statistics

noun

Statistics are a collection of data used for analysis.

eg The statistics showed the relationship between adult height and infant height.

steepness

noun

Steepness is a general term for gradient.

eg The steepness of a travel graph indicates the rate of speed.
➜ gradient, incline

stem-and-leaf diagram

noun

A stem-and-leaf diagram is used for displaying data. The 'tens' digit of each piece of data forms the stem and the 'units' digit forms the leaves. The data is then grouped in class intervals of 10.

eg The marks in the Maths test were illustrated using a stem-and-leaf diagram. Maths test marks: 22, 8, 23, 22, 17, 15, 16, 20, 29, 27, 25, 20, 30

0	8
1	5, 6, 7
2	0, 0, 2, 2, 3, 5, 7, 9
3	0

stone

noun

A stone is an imperial unit of weight still used in the UK.

eg Raisa weighed 6 stone 11 pounds.
➜ pound

straight

adjective

A straight line is a line that joins 2 points without bending.

eg Draw a straight line between A and B.

straight edge

noun

A straight edge is a ruler without markings.

eg Julia drew the base of the triangle using a straight edge.

straight-line graph $x = a^2$

noun

A straight-line graph has a general equation $y = mx + c$, where m is the gradient and c is the intercept on the y-axis. All points satisfying the equation lie on the straight line.

eg

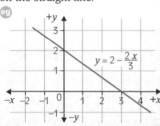

$$y = 2 - \frac{2x}{3}$$

subject of the formula

noun $x = a^2$

When a formula is arranged so that a letter equals the rest of the formula, that letter is the subject of the formula.

eg Rearrange $v = \sqrt{(u^2 + 2as)}$ to make s the subject of the formula.

substitute $x = a^2$

verb

To substitute is to exchange or replace.

eg Find the value of $v = u + at$ when you substitute $u = 8$, $a = 12$ and $t = 1$.
➜ formula

subtract [−] $+\sqrt{x} - \div$

verb

To subtract is to take away or deduct.

eg Subtract 15 from 17. What is the difference?

➡️ **deduct, minus, take away**

subtraction [−] +√x⁻÷
noun

Subtraction is the operation of finding the difference between 2 quantities.

eg Subtraction is the opposite of addition.

➡️ **decomposition**

successful
adjective

Successful means turning out in the way you had indicated or planned.

eg Probability = (number of successful events) ÷ (total number of events)

sufficient 1 2 3 4
adjective

Sufficient means enough.

eg Are the angles at the point X sufficient to make a complete revolution?

suggest
verb

To suggest is to hint or give an idea.

eg Questions in a questionnaire should not suggest an answer – that would be biased.

suit (soot)
noun

A suit is 1 of the 4 types of cards in a pack of playing cards.

eg The 4 suits are spades, hearts, diamonds and clubs.

➡️ **pack**

suitable
adjective

Suitable means appropriate.

eg A scale of 1 cm representing 5 km is suitable for this diagram.

sum
1 noun +√x⁻÷

A sum is the addition of 2 or more quantities or numbers.

eg Find the answer to the sum
$23 + 6 − 15$.

2 noun 1 2 3 4

A sum is also a quantity of money.

eg A sum of money was deposited in the bank.

supplementary angle
adjective

Two or more angles adding up to 180° are known as supplementary angles.

eg Angles 63° and 117° are supplementary as they add up to 180°.

➡️ **angles on a straight line, complementary angles, interior angle**

surface
noun

A plane space, having no thickness, is a surface.

eg The surface of a cube consists of 6 equal square faces.

surface area
noun

The surface area is the total area of the exterior surface.

eg Find the surface area of a closed cylinder with height 15 cm and diameter 8 cm.

surround
verb

To surround something is to encircle it.

eg Draw an octagon that surrounds the square.

➡️ **circumscribe**

survey
noun

A survey is a collection of data for statistical analysis.

eg The traffic survey counted all the vehicles using the crossroads.

➡ **questionnaire**

symbol
$x = a^2$

noun

A mark or sign standing for something is known as a symbol.

eg There are many mathematical symbols, including numbers (1, 2, 3, 4), signs (×, +, −, ÷) and notation ($<, \leq, =, \neq$).

symmetrical

adjective

A symmetrical figure is one that is balanced about a point, line or plane.

eg

symmetry
noun

A figure that keeps its shape if reflected or rotated is said to have symmetry.

eg Which type of symmetry do the letters A, H and Z have?

➡ **line symmetry, rotation symmetry**

system
➡ **operating system**

Tt

table
noun

A table is an arrangement of information in rows and columns.

×	1	2	3
1	1	2	3
2	2	4	6
3	3	6	9

take away [−]
verb
$+\sqrt{x}-\div$

To take away is to remove items or numbers by subtracting them.

Take away 17 from the total.
➡ **deduct, minus, subtract**

take out common factors
phrase
$x = a^2$

To take out common factors is to extract the factor common to each term.

Take out the common factors of $3y^2 - 6y$.
➡ **common factor**

tally
verb

To tally is to count by making marks.

colour of car	tally	frequency
red	JHT JHT II	12
blue	JHT II	7
white	III	3
total		22

Tally comes from the Latin word *talea*, meaning a stick. In ancient times, people kept a record of things like debts by cutting notches in a stick.

tangent [tan]
1234

1 noun

The tangent of an acute angle, in a right-angled triangle, is the ratio of the side opposite the given angle to the adjacent side.

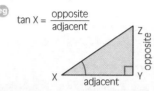

$$\tan X = \frac{\text{opposite}}{\text{adjacent}}$$

➡ **trigonometry**

2 noun

A tangent to a curve or a circle is a straight line touching it at 1 point.

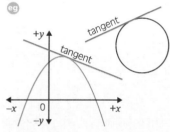

tangram
noun

A tangram is a Chinese puzzle in which a square cut into 5 triangles, 1 square and 1 parallelogram is used to make different figures.

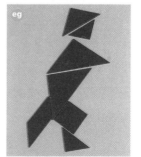

task

noun

A task is a job or assignment.

eg The task set was to design a container to hold tennis balls.

tax

1234

noun

Tax is a payment charged by the government to raise money.

eg Income tax is charged on a person's income.

➡ **value added tax**

temperature

noun

The temperature is the measure of how hot or cold something is.

eg At the beginning of January the temperature dropped 5°C.

➡ **Celsius, Fahrenheit**

tend

$x = a^2$

verb

To tend is to incline towards, or to get closer and closer to.

eg The sequence of fractions $\frac{1}{2}, \frac{2}{3}, \frac{3}{4}, \frac{4}{5} \ldots$ tends to 1.

tenth [10th, $\frac{1}{10}$]

1234

1 adjective

Tenth (10th) is the ordinal number of ten.

eg Raj was the tenth tallest pupil in the school.

2 noun

A tenth is the fraction $\frac{1}{10}$.

eg A millimetre is a tenth of a centimetre.

tera-

prefix

Tera- is a prefix meaning 1 000 000 000 000.

term

$x = a^2$

noun

A term is a part of an expression, equation or sequence.

eg What are the next 2 terms of this sequence: 5, 6, 8, 11…?

terminate

1234

verb

To terminate is to finish or end.

eg The train journey terminated at Euston station.

terminating

1234

adjective

A terminating decimal is a decimal fraction with a finite number of digits.

eg 0.75 is a terminating decimal but 0.333… is not.

➡ **recurring**

tessellation

noun

A tessellation is a pattern made by fitting together (usually regular) plane shapes without gaps.

Tessellation comes from the Latin word *tessella*, which was the name for a small piece of stone used in mosaics.

tetrahedron

noun

A tetrahedron is a solid shape with 4 triangular faces.

text

noun

Text is written material, including words displayed on a screen.

eg Word processing software enables you to type text into a computer.

theorem

noun

A statement established by proof is a theorem.

eg Pythagoras' theorem states that in a right-angled triangle, the square on the hypotenuse is equal to the sum of the squares on the other two sides.

theoretical probability

noun

Theoretical probability is predicted probability.

eg Theoretical probability is calculated using the fraction: (Number of particular outcomes that can happen) ÷ (Number of outcomes that are possible from the task).

theory

noun

A theory is a set of ideas and reasoning that explains something.

eg The theory of gravity explains why objects fall to the ground.

therefore [∴]

$x = a^2$

adverb

Therefore is another way of saying 'so' or 'this is the result'.

eg The triangle has 2 angles of 55°, therefore it must be an isosceles triangle.

thousandth

$1\,2\,3\,4$

1 adjective

Thousandth (1000th) is the ordinal number of a thousand.

eg On its thousandth throw the coin turned up heads.

2 noun

A thousandth is the fraction $\frac{1}{1000}$.

eg A metre is a thousandth of a kilometre.

three-dimensional [3D]

adjective

An object that is three-dimensional has length, breadth and height.

eg

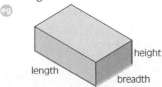

➡ two-dimensional

three-figure bearing

➡ bearing

time

noun

Time is an occasion or period, measured in seconds, minutes and hours.

eg What time does the train reach Manchester?

➡ calendar, clock

title

noun

A title is the name given to a project.

eg The title of the graph is 'A bar chart to compare heights of Year 9 pupils'.

ton (*tun*)

noun

Ton is a unit of measurement of weight or mass in the imperial system.

eg 1 ton = 2240 lb or 1016 kg

➡ tonne

tonne (*tun*)

noun

Tonne is a unit of measurement of weight or mass in the metric system.

eg 1 tonne = 1000 kg

➡ ton

toss
verb

To toss a coin is to throw it in the air and allow it to fall to the ground.

eg Toss the coin to see whether it falls heads or tails.

total
1 noun

The total of a group of numbers is their sum.

eg What is the total of 26, –13, 12 and 7?

2 adjective

The total amount is the whole quantity.

eg The total population of UK is between 55 and 60 million.

touch
verb

To touch something is to make contact with it.

eg The tangent touches the circle.

towards
preposition

To move towards something is to move in its direction.

eg The athlete is running towards the finishing line.

Tower of Hanoi
noun

The Tower of Hanoi is a mathematical puzzle.

eg

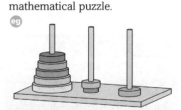

🔎 This puzzle, invented in 1883, consists of 3 pegs on a board. There are 8 circular discs of different sizes with central holes. The discs are placed on the pegs, with the largest disc at the bottom and the smallest at the top. They are then moved from peg to peg, ending with all the discs in their correct order on another peg. The idea is to count the number of moves needed for moving different numbers of discs.

transfer
$x = a^2$
verb

When you move something from one place to another, you transfer it.

eg Solve the equation $4x + 1 = 3x - 10$ by transferring the x terms to the left-hand side and the number terms to the right-hand side.

transformation
noun

A transformation is a change made to the position and/or size of a shape.

eg Enlargement, reflection, rotation and translation are all transformations.

translation
noun

Translation is a transformation in which every point of a shape moves the same distance and direction as specified by a vector.

eg

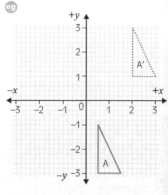

translation $\begin{pmatrix} 1.5 \\ 4 \end{pmatrix}$ maps A onto A′

transversal
noun

A transversal is a line cutting across 2 or more lines.

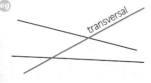

➡ intersect

trapezium (*trap-ee-zee-um*)
noun

A trapezium is a quadrilateral with 1 pair of parallel sides.

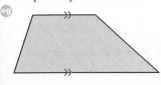

➡ isosceles trapezium

trapezoid (*trap-iz-oid*)
noun

A trapezoid is a quadrilateral with no parallel sides.

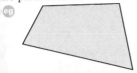

travel graph   1 2 3 4
noun

A travel graph is a diagram illustrating a journey, in which distance is plotted against time.

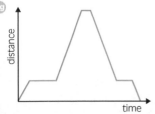

tree diagram
noun

A tree diagram is a way of illustrating probabilities in diagram form.

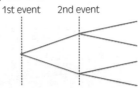

1st event   2nd event

trial
noun

A trial is a test or experiment.

The coin was tossed 100 times in a trial to test whether the coin was biased.

trial and improvement
noun $x = a^2$

Trial and improvement is a way of solving an equation by substituting values and trying to get nearer and nearer to the answer.

Solve $x^3 + x = 20$ using trial and improvement, giving the answer to 1 decimal place.

triangle
noun

A triangle is a 3-sided polygon.

Draw a triangle with 3 different angles and 3 different sides.

➡ acute triangle, equilateral triangle, isosceles triangle, obtuse triangle, right-angled triangle, scalene triangle

triangular number   1 2 3 4
noun

A triangular number is a number which can be shown by a triangular array of dots.

6 is a triangular number

triangular prism
noun

A triangular prism is a prism with a cross-section of a triangle.

eg

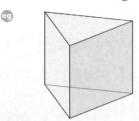

➡ prism

trigonometry
noun

Trigonometry uses the ratios of sides and angles of a right-angled triangle for calculations.

eg Use trigonometry to find an angle when you know 2 sides of a right-angled triangle.

➡ cosine, Pythagoras' theorem, sine, tangent

trillion 1 2 3 4
noun

A trillion (1 000 000 000 000) is 1 million million.

eg 1 trillion = 10^{12}

ℹ In the UK, a trillion used to be 1 million million million (10^{18}).

true
adjective

True means not false.

eg The temperature in Birmingham on Christmas Day was 32°C – true or false?

turn
1 noun

When it is your 'go' in a game or activity, it is your turn.

eg Samuel and Byron took turns at throwing the dice.

2 verb

To turn is to rotate.

eg The bicycle wheel turned through 2 revolutions.

➡ rotate

twice 1 2 3 4
adverb

Twice means 2 times.

eg Twice two is four.

➡ double

two-dimensional [2D]
adjective

A shape that is two-dimensional has length and breadth but no depth.

eg A rectangle is two-dimensional.

➡ three-dimensional

two-way table
noun

A two-way table is a table used when handling data to illustrate 2 variables.

eg

		team		
		red	blue	green
	7	16	18	16
year	8	15	17	16
	9	15	15	15

Uu

unbiased   ⊞
adjective

To be unbiased is to be fair or impartial.

eg Make sure that your questionnaire is unbiased.

➡ **biased**

uncertain   ⊞
adjective

If something is uncertain it is doubtful or unsure.

eg As the probability tends nearer to 0, the event becomes more uncertain.

➡ **certain**

underline   ? ? ?
verb

To underline something is to draw a line underneath it.

eg Remember to underline the title of the graph.

unfair   ⊞
adjective

Unfair means biased or partial.

eg That question on your survey is unfair and should be reworded.

➡ **fair**

uniform $1\,2\,3\,4$
adjective

Something that is uniform does not change, but is regular or constant.

eg A cylinder has a uniform cross-section: cut it anywhere along its length, perpendicular to its axis, and you will see a circle.

union $1\,2\,3\,4$
noun

The union of 2 or more sets of items is the sum of all the items together.

eg The union of Form 7A and Form 7B made up the whole of Year 7.

unique (*yew-neek*) $x = a^2$
adjective

Unique means being only one of a kind.

eg The unique solution to the equation $4y - 3 = 29$ is $y = 8$.

unit

1 noun $1\,2\,3\,4$

A unit means one, a single thing or number.

eg 1 unit of electricity cost 6.5 pence.

2 noun

A unit is the digit or position immediately to the left of the decimal point.

eg The units column in this sum adds up to 4:

$$\begin{array}{r} 23 \\ +41 \\ \hline 64 \end{array}$$

3 noun

A unit of measurement is a standard amount of that measurement.

eg A metre is a unit of measurement equal to 100 cm.

unitary method $1\,2\,3\,4$
noun

Unitary method is a method used in problems of proportion and ratio which involves calculating the value of 1 unit, or item, and multiplying by the number of items required.

eg Ali pays £5.32 for 7 pens. Use the unitary method to calculate how much he will pay for 5 pens.

unit fraction
noun $1\,2\,3\,4$

A unit fraction has numerator 1 and any integer as denominator.

eg $\frac{1}{5}$, $\frac{1}{11}$ and $\frac{1}{23}$ are all unit fractions.

unknown
adjective $x = a^2$

A letter in an algebraic expression or equation is called an unknown.

eg $5x + 2y + 4$ is an expression with 2 unknowns (x and y).

unlikely
adjective

If the probability of an event occurring is very small, the event is unlikely to happen.

eg It is unlikely that there will be snow in London in August.

➡ **uncertain**

upper bound
noun $x = a^2$

The upper bound is the top limit.

eg The upper bound of all numbers satisfying the inequality $1.75 \leq n < 3.5$ is 3.5.

➡ **lower bound**

upper quartile
➡ **quartile**

usual
adjective

A usual event is one which normally happens.

eg It is usual to mark the origin (the point of intersection of the x-axis and y-axis) with 0.

V v

V

1 2 3 4

V is the symbol which stands for 5 in the Roman number system.

eg C + VI + V = CXI

➡ **Roman numerals**

valid

$x = a^2$

adjective

A logical or sound argument is valid.

eg Produce a valid argument for saying that a triangle with sides 6 cm, 8 cm and 10 cm is a right-angled triangle.

value

1 2 3 4

noun

A value is a calculated amount.

eg Find the value of C in °C from the formula $C = \frac{5}{9}(F - 32)$ when F = 65°.

value added tax [VAT]

1 2 3 4

noun

Value added tax is a tax charged on goods and services.

eg A restaurant bill totals £27.50 before VAT. What is the total after VAT is charged at a rate of 17.5%?

➡ **tax**

variable

$x = a^2$

noun

A quantity that can take a range of values is a variable.

eg In the equation of a straight-line graph $y = mx + c$, both x and y are variables.

various

adjective

Various means of several different kinds.

eg Year 7 noted down various makes of cars, lorries and other vehicles.

vector

noun

A vector is a quantity with size and direction.

eg

vehicle (*vee*-ik-kl)

1 2 3 4

noun

A vehicle is a machine which conveys people or goods over a distance.

eg In town, the speed limit for all vehicles is 30 mph.

Venn diagram

1 2 3 4

noun

A Venn diagram is a way of illustrating the relationship between sets of items.

eg

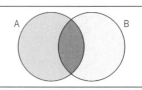

Venn diagrams are named after the mathematician John Venn (1834–1923).

verify

$x = a^2$

verb

To verify something is to confirm it or show that it is true.

eg Measure the angles to verify that angles on a straight line add up to 180°.

vertex (plural: vertices)
(**ver**-tex, **ver**-tiss-seez)
noun

A vertex is a point where 2 or more lines meet.

eg

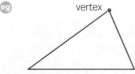

vertex

vertical
adjective

Vertical means at right angles to the horizontal.

eg The perpendicular bisector is vertical, and at right angles to the horizontal line at its midpoint.

vertically opposite angles
noun

Vertically opposite angles are formed when 2 straight lines intersect; the 4 angles add up to 360°.

eg

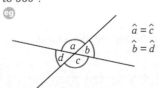

$\hat{a} = \hat{c}$
$\hat{b} = \hat{d}$

view
noun

A view of an object is the sight of that object from a given side.

eg Tina drew the rear elevation of the house, as this was the view she could see best.
➡ plan

volume
noun

Volume is the amount of space in a 3D container; it is measured in cubic millimetres, cubic centimetres and cubic metres.

eg What is the volume of a container with height 7.5 cm and a square base with side 3 cm?
➡ capacity

vulgar fraction
➡ common fraction

Ww WW Ww **Ww**

wage
noun

A wage is the money paid for work done, usually at a daily or weekly rate.

eg She collected her wages at the end of the week.

➡ **earn, salary**

week
noun

A week is a period of 7 days.

eg There are 52 weeks in 1 year.

weigh *(way)*
verb

You weigh something to find out how heavy it is.

eg The bag of sugar weighed 1 kg.

weight *(wayt)*
noun

Weight is the heaviness of an object or person.

eg Compare your weight before and after dinner.

west
noun

West is one of the 4 main points of the compass, 270° clockwise or 90° anticlockwise of north.

eg

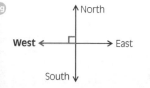

whole
adjective

The whole amount is the total quantity.

eg She used the whole can of paint to decorate her cupboard doors.

whole number
noun

A whole number is an integer, or a number used for counting.

eg A mixed number is a fraction consisting of a whole number and a fraction.

➡ **integer**

wholesale price
➡ **cost price**

wide
1 adjective

Wide means broad.

eg The road is known as Broad Lane because it is so wide.

2 adjective

Wide can also mean over a considerable range.

eg The data collected needed to be over a wide range, covering all values.

➡ **broad, narrow**

width
noun

The width of an object is the measure of how wide it is.

eg If the area of a rectangle is 65 mm² and the length is 13 mm, what is the width?

➡ **breadth**

wrong
adjective

To be wrong is to be incorrect.

eg Jenny knew the angle was wrong, because the angle sum of the triangle was less than 180°.

Xx

X   1 2 3 4

X is the symbol which stands for 10 in the Roman number system.

eg XXVII + III = XXX

➡ **Roman numerals**

x-axis $x = a^2$

noun

The horizontal axis is the x-axis, or the line $y = 0$.

eg

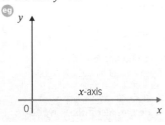

➡ **axis, y-axis**

x-coordinate $x = a^2$

noun

The x-coordinate gives the position of a point along the x-axis of a graph.

eg The x-coordinate of C(3,5,1) is 3.

➡ **coordinate, y-coordinate, z-coordinate**

yard
noun

A yard is a unit of length in the imperial system.

There are 3 feet in 1 yard.

A yard was the distance from a man's nose to the tip of his outstretched arm. The yard had to be standardised because of the different arm lengths.

y-axis $x = a^2$
noun

The vertical axis is the y-axis, or the line $x = 0$.

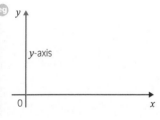

➡ axis, x-axis

y-coordinate $x = a^2$
noun

The y-coordinate gives the position of a point along the y-axis of a graph.

The y-coordinate of B(3,4,1) is 4.

➡ coordinate, x-coordinate, z-coordinate

year
noun

A year is the period of time it takes for the Earth to go round the Sun, about 365 days.

A calendar year has 12 months.

A calendar year is divided into 12 months, 52 weeks or 365 days. This definition is based on the Western or Gregorian calendar. Early calendars were lunar calendars; the lunar year has 13 months, each having 28 days. Different cultures have their own calendars, including the Chinese, Jewish and Muslim calendars.

➡ calendar, month

Z z

z-axis
noun

$x = a^2$

The z-axis gives the 3rd dimension on a graph.

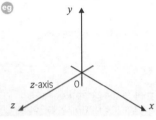

z-coordinate
noun

$x = a^2$

The z-coordinate gives the position of a point along the z-axis of a graph.

The z-coordinate of A(2,3,5) is 5.

➡ **coordinate, x-coordinate, y-coordinate**

zero [0]
noun

1 2 3 4

Zero is nothing, nought, nil.

Anything multiplied by zero equals zero; any number to the power zero equals one.

Zero was first used by people in ancient India.

zero place holder
noun

1 2 3 4

Zero place holder is a term used in the place value system.

In the number 1203 the digit 0 is a zero place holder.

A B C D E F G H I J K L M N O P Q R S T U V W X Y Z